STEREOCHEMISTRY

The Static Principles

J. GRUNDY
B.Sc., A.R.I.C.

Lecturer in Organic Chemistry, Brunel College, London

Springer Science+Business Media, LLC

1964

Suggested U.D.C. number 541.63

ISBN 978-1-4899-5897-6   ISBN 978-1-4899-5895-2 (eBook)
DOI 10.1007/978-1-4899-5895-2

© Springer Science+Business Media New York 1964

Originally published byButterworth & Co. (Publishers) Ltd. in 1964.
Softcover reprint of the hardcover 1st edition 1964

STEREOCHEMISTRY

The Static Principles

U.S.A.:	BUTTERWORTH INC. WASHINGTON, D.C.: 7235 Wisconsin Avenue, 14
ENGLAND:	BUTTERWORTH & CO. (PUBLISHERS) LTD. LONDON: 88 Kingsway, W.C.2
AUSTRALIA:	BUTTERWORTH & CO. (AUSTRALIA) LTD. SYDNEY: 6/8 O'Connell Street MELBOURNE: 473 Bourke Street BRISBANE: 240 Queen Street
CANADA:	BUTTERWORTH & CO. (CANADA) LTD. TORONTO: 1367 Danforth Avenue, 6
NEW ZEALAND:	BUTTERWORTH & CO. (NEW ZEALAND) LTD. WELLINGTON: 49/51 Ballance Street AUCKLAND: 35 High Street
SOUTH AFRICA:	BUTTERWORTH & CO. (SOUTH AFRICA) LTD. DURBAN: 33/35 Beach Grove

Da mihi, Domine, scire quod sciendum est

CONTENTS

PREFACE

Theoretical organic chemistry is conveniently divisible into its electronic and stereochemical aspects. The former focuses attention on the electronic distribution in molecules and disregards the nuclei while the latter is primarily concerned with the nuclear distribution. This division is artificial since the electronic distribution in a molecule is a stereochemical feature and is frequently important as such.

Stereochemical theory (like other aspects of chemistry) has developed enormously during the last fifteen years and now requires subdivision into 'static' and 'dynamic' stereochemistry. Static stereochemistry concerns the geometry or geometries which molecules of a particular compound may have under (usually) normal conditions; dynamic stereochemistry concerns the effect of geometry when molecules are undergoing or attempting to undergo chemical reaction. The static theory is of primary importance since not only is it sufficient in discussing many physico-chemical and bio-chemical phenomena but it is a necessary preliminary to the understanding of the dynamic aspects of stereochemistry.

It will be clear that modern stereochemical theory has assumed the status of a *sine qua non* for the practice of organic chemistry. Unfortunately, it is the author's experience that in certain quarters it has not been recognized adequately as yet that the days have passed when a 'van't Hoff treatment' of the tartaric acids together with a brief description of the diphenyls can be regarded as constituting a satisfactory knowledge of stereochemistry. This situation is no doubt due in part to the advances in knowledge but the knowledge is now of fundamental importance and the situation described requires a remedy. It is with this in mind that this book has been written; this first volume is a modern presentation of the principles of static stereochemistry and provides the basic theoretical knowledge necessary to the understanding and use of stereochemical principles in organic chemistry. The book throws the subject as a whole into modern perspective by treating the particular phenomenon of stereoisomerism as a direct derivative of the principles of molecular geometry. It is hoped that besides students of chemistry the importance of the subject matter will be recognized by those engaged in related studies.

Eastcote, Middlesex   J. GRUNDY
April, 1964

THE FACTORS AFFECTING ATOMIC AND MOLECULAR GEOMETRY:
(I) THE ATOMIC STATE, RESONANCE AND SUBSTITUENT FACTORS

INTRODUCTION

Stereochemistry concerns the geometry of atoms and molecules; in antiquity Plato (400 B.C.) regarded the atoms of Democritus (425 B.C.) as having the shapes of the five regular solids. In the modern period the development of the Daltonian atomic and molecular concepts (1808) led to the classical theory of valency (Kekulé[1], Couper[2]) and, in turn, led to the stereochemical theories of Van't Hoff[3a,b,c] and Le Bel[4]. The classical stereochemical theory was built on the experimental data of organic chemistry and naturally the geometry of the carbon atom was the primary concern; this atom was envisaged as having its four equivalent valencies disposed tetrahedrally in space. A simple pictorial model represented the valencies as lines radiating tetrahedrally from the atom's symbol. Stereochemical thinking was extended by Werner[5] to certain inorganic compounds (co-ordination compounds) and by 1911, on the basis of chemical evidence, specific geometries had been suggested for the valency distributions of the atoms of nitrogen, sulphur, selenium, tin, silicon, phosphorus and arsenic in certain of their compounds.

The actual geometry of the valency distributions of many atoms is now known as a result of the more recent development of physical methods (e.g. x-ray, dipole moment methods, etc. 1920 et seq). These methods have also provided extensive data on valency angles and bond lengths so giving a more definitive picture of atomic and molecular geometry than was possible in the classical period. The physical measurements have provided the important generalizations that the valency distribution of a particular atom in various molecules remains essentially constant and that the geometries of atoms conform to a quite limited number of types.

The development of the classical electronic theory of valency made possible some correlation of the geometrical types with the electronic structure of atoms in molecules. Organic molecules contain atoms (C, N, O, P, S and halogens) which usually have an octet of electrons in their outer shell when in combination in molecules; these octet atoms may be, of course, mono-, di, tri- or tetracovalent. A monovalent atom is clearly of no stereochemical interest but a dicovalent atom (O, S) has its valencies at an angle of between 90 degrees and 110 degrees and generally at about 105 degrees. A trivalent octet species (N, O, P, S) has a pyramidal valency distribution and the angle is again between 90 degrees and 110 degrees, dependent on the particular molecule. The tetracovalent type of atom includes nitrogen and phosphorus atoms as well as the carbon atom; there is some evidence that oxygen atoms can act tetracovalently and although the sulphur atom can have four groups attached by electron pair covalencies it is now known that other electrons can simultaneously participate in bonding and double bonds can exist. Nevertheless, in octet compounds involving these atoms the electron pairs whether used in bonding or present as lone pairs are distributed tetrahedrally or essentially so and therefore have the valency geometry postulated by the classical theory for the carbon atom; when such an atom is dicovalent or tricovalent an angular or pyramidal molecule will result (Sidgwick and Powell[6]).

The above correlations of geometry and electronic structure, while useful, were applied earlier to saturated molecules, thus, the tetracovalent carbon atom is tetrahedral in saturated aliphatic molecules but it has a quite different geometry in ethylenic and acetylenic molecules. These different geometries for a particular atom could not be interpreted by classical electronic theory; they arise because the pairs of electrons of the octet can be distributed in different ways amongst the available molecular energy levels; each mode of distribution corresponds to a different geometry of the valencies. The understanding of these matters and, therefore, of the stereochemistry of atoms and molecules has been made possible only by the development of the modern quantum mechanical atomic and valency theories (1926 et seq).

As pointed out, an atom in a particular electronic state has a particular distribution of its covalencies; the directional properties of the covalent bonds of particular atomic states therefore, are principally responsible for the geometry of molecules. However, even for given states of the atoms in molecules the molecular geometry does not follow entirely since joining atoms together invokes other factors which influence the resultant molecular geometry; these various factors may be discussed now in detail.

THE ATOMIC STATE FACTOR

The Carbon Atom

The Tetracovalent Carbon Atom

The classical Kekulé–Couper theory regarded the carbon atom as invariably tetravalent and the entire classical stereochemistry of organic molecules was built very successfully on the concept of a tetrahedral distribution of the four valencies. The development of physical methods and of quantum mechanics have substantiated the tetrahedral model for aliphatic molecules but have shown also that the tetracovalent carbon atom can exist in several electronic states each having its characteristic geometry.

The sp^3 or tetrahedral carbon atom—The gross electronic configuration of the carbon atom is $1s^2 \cdot 2s^2 \cdot 2p^2$. (Bohr[7], Stoner[8]), however the three p orbitals which make up the $2p$ sub-shell have the same energy ('degenerate') and therefore, since electrons maximally uncouple (Hund[9]) the detailed configuration of the ground state is $1s^2 \cdot 2s^2 \cdot 2p \cdot 2p$. and the uncoupled electrons have parallel spins. Organic molecules are held together largely by covalent bonds and modern quantum theory concurs with classical electronic theory (Lewis[10]) that such bonds are made up of a pair of electrons. This electron pair usually derives from a one electron contribution by the two bonded atoms and the Pauli[11] principle (1925) requires that the bonding pair have opposed spins. The electrons used by an atom for bond formation are peripheral ones since these are of the highest energy. These conclusions applied to the ground state carbon atom suggest that it should be divalent since it has two unpaired peripheral electrons but, of course, the atom is tetra-covalent. This problem is surmounted theoretically by assuming that an electron excitation or promotion can occur from the $2s$ sub-shell to the vacant $2p$ orbital; this process requires 90–100 kcal/g-atom and results in the configuration $1s^2 \cdot 2s \cdot 2p \cdot 2p \cdot 2p$. This excited atom can provide four covalencies because of its four unpaired electrons but, the s and p orbitals are not identical and therefore one of the four covalencies will differ from the other three.

An orbital is defined as a region of extra-nuclear space in which an electron may move; atomic orbitals may be regarded roughly as the quantum mechanical equivalents of Bohr orbits. However, in 'orbitals', unlike 'orbits', the electronic motion is not exactly predictable, i.e. is non-Newtonian, and all that can be said is that the probability of finding an electron at any time and at any point of extra-nuclear space has a certain value. The probability of an

3

electron's occurrence varies from point to point of extra-nuclear space; the probability in the vicinity of the nucleus and at large distances from the nucleus is low but between these points the probability rises to a maximum value at a distance from the nucleus which depends on the electronic energy level. The probability concept of electronic motion may be represented pictorially (Schrödinger[12]) by imagining the electron to be smeared out over extra-nuclear space as a cloud of charge so that the density of the charge cloud at any point of space is proportional to the probability of the electron's occurrence at that point; this picture would correspond to a time exposure photograph of the electron if this was obtainable. The pictorial usefulness of this representation is completed by drawing a boundary surface enclosing an arbitrary amount (90 per cent) of the charge cloud. These pictorial methods and the mathematics of the quantum theory show that s orbitals are spherically symmetrical about the nucleus whereas the three p orbitals of a sub-shell are 'dumb-bell' shaped and are mutually perpendicular. The charge clouds of s and p orbitals have the shape of the solids of revolution generated by rotating *Figure 1a* and *b* respectively about the axes shown; as pointed out three such p orbitals at right angles to each other exist and with their 'dumb-bell' centres coincident.

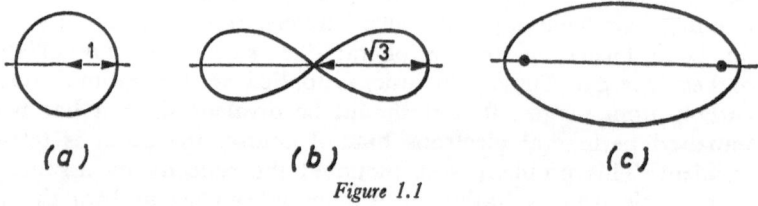

(a)   *(b)*   *(c)*

Figure 1.1

The formation of a covalent bond may be envisaged (Pauling[13b,c]) as a penetration or overlap of the two orbitals of the atoms concerned; each orbital contains an electron and orbital overlap results in a build-up of electronic charge between the nuclei. This increased charged density corresponds to a high probability of finding the electron pair between the nuclei and this quantum mechanical result is closely similar to the classical Lewis picture with the important difference that classical theory held that the electrons are static. This simple geometrical overlap idea is only pictorial and does not represent the actual resultant picture since the bonding electrons move in the field of the two bonded nuclei and therefore the geometry of the component atomic orbitals is changed somewhat; electrons moving about two or more nuclei are said to occupy a 'molecular orbital'. Thus, when the two $1s$ orbitals of two hydrogen atoms overlap to form a bonding orbital the resultant molecular orbital has the shape represented by *Figure 1.1c*. The charge distribution in a molecular orbital is such that there is a concentration of charge between the bonded nuclei and this binds the nuclei together by electrostatic forces. In the $1s$ state of the hydrogen atom the energy is made up of the kinetic and potential energies of the electron; when a hydrogen molecule is formed the energy of the system is made up of the potential and kinetic energies of the electrons together with the energy of nuclear repulsion. In the molecule electrons move over both nuclei and while this has little effect on their kinetic energies it results in considerable decrease in potential energy relative to the two isolated atoms from which the molecule derives. This change in potential energy more than offsets nuclear repulsion and a stable system results; this is the source of bond energy. In terms of orbital overlap the optimum concentration of charge between bonded nuclei requires 'maximum overlap' of the orbitals and this results in a specific direction of overlap where this is possible. Thus the principle of maximum overlap of orbitals in bond formation clearly has geometrical consequences and indeed is fundamental to the interpretation of the directed covalent bond. In the formation of a hydrogen molecule from ground state atoms the two $1s$ orbitals may approach and overlap from any direction because of the spherical symmetry of the orbitals, however, if an excited molecule is formed from two atoms in their $2p$ states the overlapping takes place along the axes of maximum extent; this maximum overlapping is described as 'linear overlap'.

The $2p$ orbitals of an atom extend $\sqrt{3}$ units in space and have a specific directional character (*Figure 1.1b*) whereas the $2s$ orbital extends relatively one unit and being spherically symmetrical has no specific directional character (*Figure 1.1a*). The spin uncoupled carbon atom $1s^2 \cdot 2s \cdot 2p \cdot 2p \cdot 2p$. should form, using p orbitals, three mutually perpendicular bonds with other atoms together with a fourth bond, from the $2s$ orbital, whose orientation will be determined by the presence of p bonds and not by any specific directional properties of s orbitals. However, the long established equivalence of the four valencies of carbon (e.g. in methane CH_4, Henry[14]) and their tetrahedral distribution (van't Hoff[3a], Le Bel[4]) does not accord with the above quantum mechanical model. This problem is solved by invoking a further quantum mechanical principle of the greatest general importance called the principle of 'orbital hybridization'

(Pauling). The principle of hybridization states that, in considering the overall process of molecule formation from atoms, it is sometimes advantageous to consider the 'mixing' or hybridization of energetically similar atomic orbitals. The mixing of the orbitals of an atom, a mathematical manipulation, results in the formation of new orbitals having different geometries to the atomic orbitals from which they derive; the basis of the hybridization concept is that the new hybrid orbitals are superior to the original orbitals since they permit the formation of stronger bonds. The mathematics allows all degrees of 'mixing' but the chemistry of the atom limits the possibilities, thus, for the aliphatic carbon atom the four bonds are equivalent and so the 'mixing' process must provide four new equivalent orbitals from the $2s$ and the three $2p$ orbitals; the hybrid orbitals clearly have 25 per cent s and 75 per cent p character; the calculations show (Pauling[13a,c]) that four such hybrids (sp^3 hybrids) are directed towards the corners of a regular tetrahedron and have an axial extent of 2 units (*Figure 1.2a*). The conversion of $1s^2 . 2s^2 . 2p . 2p$ carbon via $1s^2 . 2s . 2p . 2p . 2p$. to the hybridized state requires 90–100 kcal/g-atom for each step. However, as pointed out, the sp^3 hybrids are superior for bond formation and the energy furnished during bond formation with atoms more than compensates for that outlayed to produce the sp^3 state.

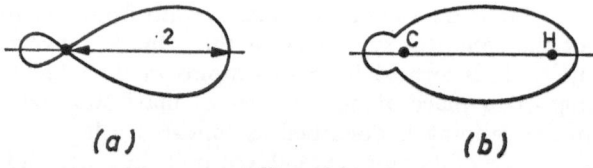

(a)   (b)

Figure 1.2

Molecule formation using sp^3 carbon gives a polyatomic molecule and such molecules involve another quantum mechanical principle of stereochemical importance, namely, the concept of the 'localized molecular orbital.' Thus, in the formation of the polyatomic molecule CH_4, from an sp^3 carbon and four hydrogen atoms, each sp^3 hybrid overlaps maximally with an s orbital of a hydrogen atom to form a molecular orbital of the shape shown in *Figure 1.2b*. These four molecular orbitals overlap in the region of the carbon nucleus and, on general theory, it is expected that the electrons can move over all five nuclei of the molecule. However, in terms of the localized orbital concept, each sp^3-s orbital is considered to be bicentric and a pair of electrons moves over the carbon nucleus

6

and the associated hydrogen nucleus. Thus, in localized terms, the bonds of a polyatomic molecule each resemble a bond of the type discussed for the hydrogen molecule. The localized molecular orbital clearly has the same general geometry as the original sp^3 atomic orbitals and this permits the general concept of tetrahedral carbon. Thus, localized molecular orbitals are pictorially conceived as deriving geometrically from their atomic orbitals and this can be envisaged whatever the nature of the groups attached to say, an sp^3 carbon atom. Non-localized orbitals abolish this generality and the existence of tetrahedral carbon requires separate mathematical derivation for each molecular species. A further advantage of localized orbitals is that the representation of a bond by a pair of electrons largely resident between the nuclei is retained. The concept of the localized orbital is not to be confused with the problem of classical electronic theory which could not reconcile the concept of directed valency with the dynamic nature of electrons; the modern theory uses the dynamic property of electrons but limits their motion in space to a localized orbital. The above discussion of sp^3 carbon is interpretative; the sp^3 concept is a mathematical description of the tetrahedral valency distribution of the carbon atom in saturated molecules; isolated carbon atoms do not exist in the sp^3 state.

One concludes that quantum mechanical theory advances the van't Hoff theory by accounting for the four tetrahedral covalencies of the aliphatic carbon atom (sp^3 carbon). The theory is based on the concepts of 'maximum overlap', the 'localized molecular orbital' and the 'hybridization' of atomic orbitals.

The sp^2 or planar carbon atom—The doubly bonded carbon atom was introduced by Erlenmeyer[15] to represent the ethylenic unit $>C = C<$; this formulation retains the fundamental tetravalent concept for carbon (Kekulé, Couper). The geometry of the ethylenic unit was represented by van't Hoff[3b] as two linked tetrahedral carbon atoms.

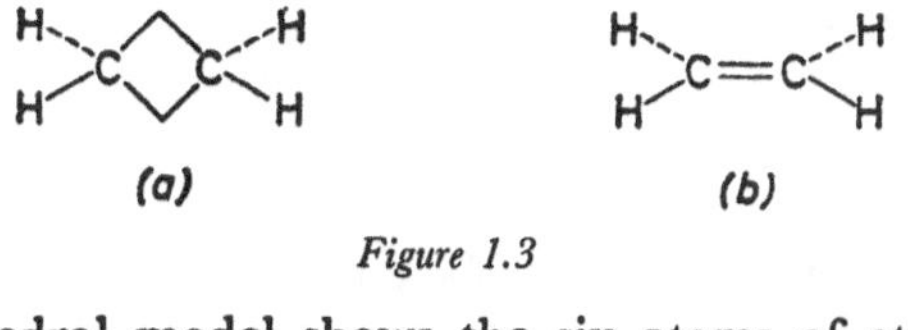

(a)   *(b)*

Figure 1.3

This tetrahedral model shows the six atoms of ethylene to be coplanar and the two angular bonds lie in a plane perpendicular to the six atom plane (*Figure 1.3a*).

The classical theorists intuitively felt that straight lines between atoms more correctly represent the properties of valencies and so Baeyer[16] constructed the double bond by bending two valencies of a tetrahedral carbon atom through a half of the tetrahedral angle (54° 44′) and then joining two such atoms (*Figure 1.3b*). The Baeyer theory was elaborated by Ingold and Thorpe[17] by considering how the strain in the Baeyer model is best distributed; it was suggested that the strain can be taken up both by the valencies and by the carbon atoms thus the valencies occupy curved paths instead of the straight paths of the Baeyer model. This Baeyer–Ingold–Thorpe model has the same general geometry as the van't Hoff model excepting for the bent bonds; it has however the additional ability to interpret the reactivity of the ethylenic unit in terms of molecular strain. Physical methods have confirmed the planarity of the ethylenic unit but show that the valency angles are 120 degrees and not tetrahedral ones of 109 degrees 28 minutes, further, details of bond lengths are known whereas classical theory could say nothing about such fine structures. It is to be realized that while physical methods can determine bond angles and lengths they cannot ascertain the nature of the double bond or determine whether one even exists; this is a theoretical problem.

The absence of tetrahedral angles in the ethylenic unit clearly invalidates a molecule based simply on sp^3 hybrids and a further reason is that the use of sp^3 orbitals allows only non-maximum, 'angular overlap' and, as will be later, shown such 'bent' or 'banana' bonds are very much weaker than those obtaining from maximum linear overlap. However, the use of 'bent' bonds in the construction of molecules is not excluded and indeed the idea is applicable to the carbon-carbon double bond although the 'bent' bonds are not overlapping sp^3 hybrids. The quantum mechanics envisages the ethylenic unit as constructed from a second type of hybrid orbital called the sp^2 orbital. This type of orbital derives from the $1s^2 . 2s . 2p . 2p . 2p .$ atom by the 'mixing' of the $2s$ and two of the $2p$ orbitals. The hybridization of these orbitals gives three equivalent coplanar sp^2 hybrids having angles with each other of 120 degrees and having axial extent of 1·991 units (*Figure 1.4a*). The remaining $2p$ electron occupies its pure $2p$ orbital and this is orientated perpendicular to the sp^2 hybrid orbital plane (*Figure 1.4b*). The carbon-carbon double bond is established by joining two such atoms; a bond is established by the linear overlap of an sp^2 hybrid from each atom and the second bond is established by the 'lateral overlap' of the two pure p orbitals (Penney[18] and Hückel[19]). The linear sp^2—sp^2

overlap bond is the stronger and is called a σ-bond; lateral overlap gives a weaker bond and this is called a π-bond. This quantum mechanical picture therefore shows the two bonds of a carbon-carbon double bond to be rather different structures but, of course, the classical theory was unable to show this. The σ-π-model clearly agrees with the established geometry of the ethylenic unit.

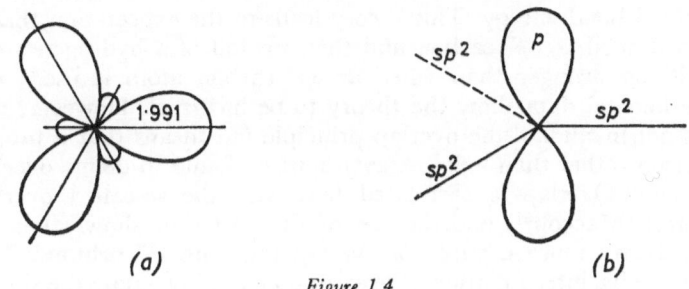

(a)   (b)

Figure 1.4

The overlap of orbitals in the σ-π-model produces a σ- and a π-molecular orbital of different detailed geometry to the component atomic orbitals; the formation and general geometry of a π-molecular orbital is shown in *Figures 1.5a, b* and *c*. The region of space close to the atomic plane has zero density of electronic charge;

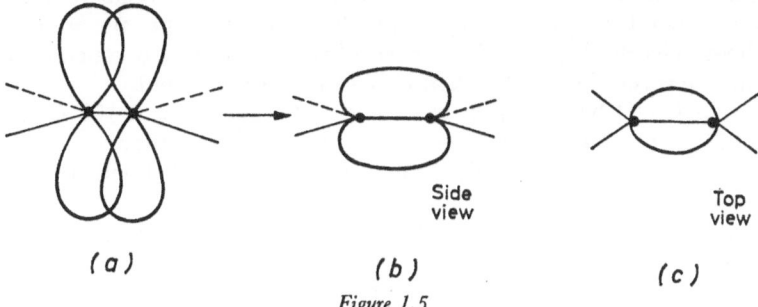

(a)   (b)   (c)

Figure 1.5

such regions are called nodes. The existence of a node might appear to prevent the motion of an electron from one part of the orbital to the other, in classical mechanics this would be true but in quantum mechanics it is not. One concludes that the quantum mechanical theory in the form of the localized orbital concept, pictorially envisaged in terms of atomic orbital overlap, accounts for the geometry of a carbon-carbon double bond.

In addition to the general geometry of molecules the modern theory can deal with certain other matters such as bond energies. The bond energies for the C—H bonds of ethane and ethylene are about 103 and 106 kcal/mole respectively. An interpretation due to Pauling and Sherman[20] envisages bond energy to be related to the overlap capacity of the atomic orbitals concerned, thus, the greater the axial extent of the atomic orbitals the greater the expected bond energy. This theory leads to the expectation that a bond formed by sp^3 carbon and the s orbital of a hydrogen atom would be stronger than when an sp^2 carbon atom is used; the experimental data show the theory to be incorrect. However, this does not invalidate the overlap principle but means that a proper criterion (other than axial extent) must be found to define overlap capacity. Overlap is correlated best with the so-called overlap integral (Maccoll[21]) and the use of this criterion shows that sp^2 hybrids are more capable of overlap than are sp^3 orbitals. This quantitative interpretation of overlap does not invalidate the simple qualitative use of the idea in connection with matters such as σ- and π-bonds; lateral overlap will give a weaker bond. The carbon-carbon bond energies in ethane and ethylene are 83 and 146 kcal/mole thus as an approximation the π-bond has an energy of 63 kcal/mole and although weaker than the σ-bond it cannot be regarded as a weak bond. The reactivity of this bond appears only because the addition of two atoms to a double bond replaces the π-bond by two strong σ-bonds. The quantum mechanical theory clearly disposes of the need of the Baeyer strain concept for the interpretation of the chemistry of the double bond.

The C—H bond lengths in ethane and ethylene are 1·107 Å and 1·071 Å respectively; the C—C and C=C bond lengths are 1·536 Å and 1·353 Å respectively. The interpretation of bond lengths involves various factors and cannot be done satisfactorily as yet. However, the shortness of the C=C relative to the C—C is expected since the concentration of the charge of four electrons between the nuclei of the carbon-carbon double bond will draw the nuclei more closely together than in a carbon-carbon single bond.

The classical theory of stereochemistry is based on the tetrahedral carbon atom, however, for reasons to be discussed later two corollaries to the tetrahedral hypothesis are necessary. The first of these is called the principle of 'free rotation' (van't Hoff[3c]) and this states that parts of a molecule can rotate freely about a single bond within a molecule; thus, a methyl group of ethane can rotate about the carbon-carbon bond. This principle is interpreted easily in quantum mechanical terms; thus in the ethane molecule and thinking of the carbon-carbon bond in terms of the overlap of two atomic orbitals, it is clear that a rotation of the CH_3 group about this bond axis cannot affect the degree of overlap of the two sp^3 orbitals making up the bond and rotation is not prevented. The second principle, also due to van't Hoff[3c], states that rotation of parts of a molecule about a double bond does not normally occur because of the rigidity of the bond. This classical principle follows from a mechanical tetrahedral model of the double bond; a rotation is impossible without breaking bonds. The rigidity of a double bond follows from the σ-π-model; thus the geometry of p orbitals results in maximum overlap to form a π-bond only when the orbitals are parallel to each other. A rotation of parts of the molecule so that the p orbitals are out of parallel decreases the overlap and weakens the π-bond. A rotation of the p orbitals to an orientation of 90 degrees to each other results in no significant overlap and the π-bond is effectively broken. In view of the energy of the π-bond the rotation about the double bond is resisted so accounting for its rigidity.

The sp or linear carbon atom—A third, long recognized state of the carbon atom is when it forms part of a triply bonded or acetylenic unit. The problem of the —C≡C— unit is similar to that of the >C=C< unit both on the classical and the quantum mechanical theory. The van't Hoff picture of the triple bond is of two tetrahedral carbon atoms linked together by three valencies from each; on this model the atoms of acetylene HC≡CH are collinear. The Baeyer geometry is also linear; each carbon atom contributes three strained valencies which are joined to form the triple bond. These models agree with the physically established linear geometry of the triple bond and the Baeyer model provides a strain interpretation of reactivity, indeed the strain theory originated from Baeyer's experience of the explosive reactivity of certain acetylenic compounds.

A quantum mechanical model based purely on sp^3 hybrids is invalid in view of the energetics of 'bent' bonds and also because the C—H bond length is not the same in acetylene as it is in ethane. The generally accepted quantum mechanical theory of the carbon-carbon triple bond uses carbon atoms which form two sp hybrid orbitals. The theory shows that when two sp hybrids are formed from a $1s^2 . 2s . 2p . 2p . 2p .$ carbon atom they are oriented at 180 degrees to each other; the remaining two pure p orbitals are oriented at 90 degrees to each other and to the two sp hybrids

(*Figure 1.6a*). The C≡C unit arises by the linear overlap of two *sp* hybrids, one from each carbon atom, together with the lateral overlap of the two pairs of pure *p* orbitals. The *sp—sp* overlap provides a σ-bond and the *p—p* overlaps provide two π-bonds (*Figure 1.6b*). The energies of the C—H and C≡C bonds in

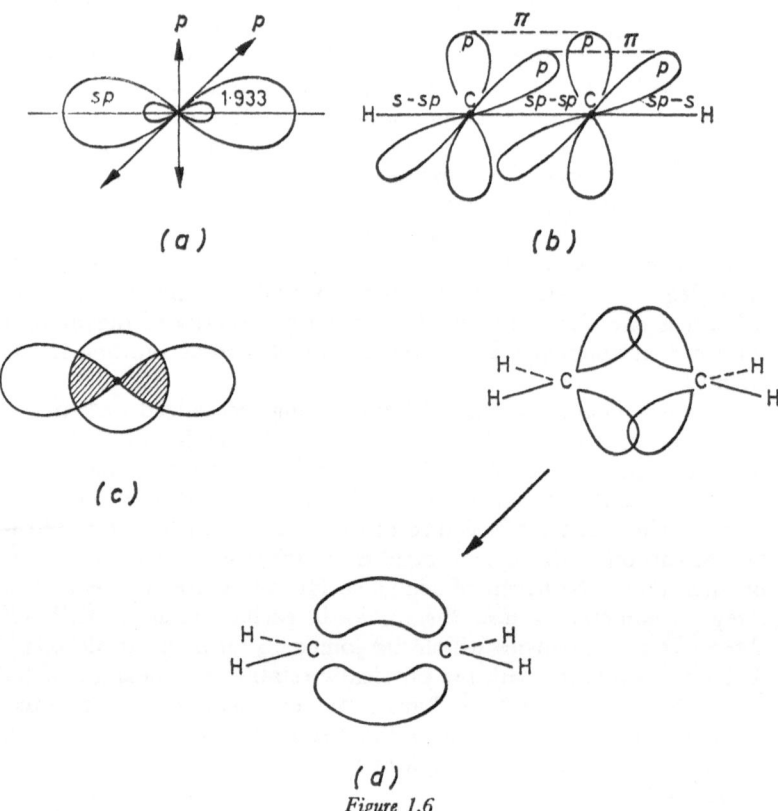

(a)

(b)

(c)

(d)

Figure 1.6

acetylene are 121 and 201 kcal/mole respectively and the overlap integral accounts for these values satisfactorily; the axial extent of *sp* hybrids is 1·933 units. The C—H bond length is 1·063 Å and that of the C≡C is 1·201 Å; these lengths cannot be accounted for in detail but the shortening of the C≡C length relative to the C=C length is expected because more electrons are concentrated between the nuclei. The theory shows that the ethylenic and acetylenic units have a general structural similarity, however, differences of detail do exist. Thus, in the $C\equiv C$ unit the two π-molecular orbitals result in complete symmetry of the electron cloud about the carbon-carbon bond axis and clearly this permits free rotation about the triple bond; classical theory, as a mechanical model shows, considered triple bonds to be rigid structures. The chemistry of the $C=C$ and $C\equiv C$ is similar but important differences exist which are traceable to the fact that in the triple bond the π-electrons are less polarizable and more localized centrally between the nuclei (Koch and Hammond[22]).

The stereochemical conclusions relating to the sp^3, sp^2 and sp states of the carbon atom strictly apply to the hypothetical case of the isolated atoms. However, the physical data shows that the stereochemistry of the sp^3, sp^2 and sp states is simulated closely in the simple ethane, ethylene and acetylene molecules. The various types of hybrid are characteristic of aliphatic, ethylenic and acetylenic systems because they are the most stable hybrids for these systems. Thus, a molecule of ethane may be derived theoretically from ethylene by breaking the π-bond and then attaching two hydrogen atoms to the available p orbitals. This does not happen, four equivalent sp^3 orbitals are used by the carbon atoms in ethane and clearly the sp^3 state is the most stable one for saturated carbon; this result was intuitively employed by Baeyer.

The above treatment of the carbon atom has considered the geometry of the various states in terms of individual forms of hybridization. An alternative quantum mechanical analysis, related to the earlier ideas of Sidgwick and Powell, provides a more general approach to the interpretation of the stereochemistry of atoms. This method is based on the principles that (a) electrons repel each other because of their charge and so have a low probability of being found close together in extranuclear space ('charge correlation') (b) electrons of the same spin repel each other ('spin correlation') (c) electrons of opposed spin experience only charge correlation effects but these are of insufficient magnitude to prevent such electrons pairing together.

The four valency electrons of carbon were described previously as distributed in the s and p orbitals. However, these orbitals overlap considerably (*Figure 1.6c*) and so the operation of spin and charge correlation will result in a low probability of finding the electrons in the region of overlap. It follows that this distribution description is inadequate and an analysis in terms of correlation effects shows that the most probable arrangement of the four electrons is at the apices of a regular tetrahedron (Zimmerman and van Rysselberghe).[23]

This tetrahedral distribution caused by correlation effects may be described by invoking the sp^3 orbital concept.

Molecule formation is best envisaged by using C^{4-}, rather than the carbon atom, as the starting system. The four pairs of electrons of the ion are also distributed tetrahedrally due to correlation effects. The molecule is completed by bringing up the appropriate particles (e.g. H^+ in CH_4) to the C^{4-} ion so that the electron pairs form bonds. The general application of these ideas to the geometry of molecules and ions may be made by invoking the further rule that repulsion between the electron pairs of a valency shell decreases in the order, lone pair-lone pair $>$ lone pair-bonding pair $>$ bonding pair-bonding pair. The C^{4-} ion has its four pairs of electrons distributed tetrahedrally and at a particular distance from the nucleus; if four protons are brought up to bond then the electron pairs will be drawn out slightly further from the nucleus. However, since the protons are identical, this effect will be the same for each of the electron pairs and the tetrahedral geometry will be maintained. The carbon ion $\bar{C}$ may be derived from CH_4 by removal of a proton, the lone pair so generated will more repel the three bonding pairs and their tetrahedral angle will decrease. This ion will have therefore an irregular tetrahedral geometry. The carbonium ion is not an octet species though correlation effects apply equally. Thus the most probable distribution of three electron pairs about a nucleus can be shown to be at the vertices of an equilateral triangle. This is precisely the description of this ion which was arrived at previously. The geometry of multiply bonded systems can be derived also from the tetrahedral model. Thus, starting with the four tetrahedrally distributed pairs of electrons, two pairs of which are bonded to hydrogen, one imagines a second carbon atom to be brought up to bond with the remaining two pairs. A strong interaction will result and the two pairs of electrons will be drawn towards the carbon-carbon axis so forming the double bond. The tetrahedral angle of the double bond electrons will be reduced as a consequence. It follows that as the electrons associated with the double bond come under the influence of the second carbon atom they will interact much less with the electron pairs of the C—H bonds. These pairs will then be subject to increased repulsion between themselves and angle widening to 120 degrees occurs. A model of the double bond on this theory is shown in *Figure 1.6d*; the electrons in their initial sp^3 orbitals being drawn towards the C—C axis are described as occupying 'bent' or 'banana' orbitals in the formed double bond. This model is clearly a better description than the $\sigma-\pi$ one since in the latter considerable overlap of the σ- and π-orbitals occurs. The 'bent' bond model of the double bond is clearly the quantum mechanical equivalent of the classical Baeyer–Thorpe–Ingold model. It should be pointed out that even when π-bonds are invoked to describe a molecule the correlation theory is very useful and permits the deduction of geometry without direct use of a particular hybridization device. Thus the carbon atoms of ethylene each have three atoms attached hence they must have three σ-pairs. It is σ-pairs only which need be regarded as determining stereochemistry. As pointed out in connection with the $\overset{+}{C}$ ion three pairs of electrons are disposed at the vertices of an equilateral triangle and the recognition of this suffices to interpret the geometry of ethylene. The π-bond is added as a stereochemically unimportant afterthought to complete the molecule. Both the correlation and 'classical' hybridization methods will be used in parts of subsequent discussion for describing molecular geometry.

Other hybrid states of the carbon atom—As indicated previously the sp^3, sp^2 and sp states of the carbon atom are idealized states. Thus, the greater the differences in the natures of substituents attached to say an aliphatic saturated carbon atom the greater will be the irregularity of the tetrahedral distribution of the valencies and the greater the deviation, from sp^3, of the carbon orbitals used. This possibility was appreciated in the classical period and both van't Hoff and Le Bel allowed that deviation from the regular tetrahedral might occur when the groups attached to the carbon atom are different. However, as physical methods show, these differences are usually small and saturated carbon atoms, independent of the molecule, have their valencies near the tetrahedral and their orbitals are, therefore, essentially sp^3. This conclusion applies to most molecules but important exceptions are known in which a tetracovalent carbon atom has a geometry which deviates from the sp^3, sp^2 and sp types; the molecules of cyclopropane and cyclobutane are examples. The cyclopropane molecule (I) has its carbon atoms at

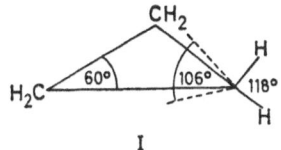

I

the vertices of an equilateral triangle and has $\overset{\frown}{HCH} = 118$ degrees (Bastiansen and Hassel[24]). The molecule clearly cannot be constructed from sp^3 carbon atoms which overlap linearly. Hybrids making an angle of <90 degrees with each other are impossible hence the bonds cannot be pure σ-bonds. The problem may be considered by starting with sp^3 carbon atoms; these orbitals can overlap in only an angular manner and the 'bent' bonds will make the molecule very unstable. The molecule may be made more stable by using hybrids other than sp^3; this will decrease the hybrid angle and permit more linear overlap of the orbitals. However, the sp^3 state is optimally stable and while a change from sp^3 will decrease the 'bentness' of the bonds and stabilize the molecule a change from sp^3 is a destabilization. A compromise between bond bending and the variation from sp^3 is reached and the orbitals used are $sp^{4\cdot12}$; the angle between the hybrids is 106 degrees (Coulson and Moffitt[25]). The presence of 'bent' bonds in molecules results in shorter bond lengths (internuclear distances); the C—C length in cyclopropane is $1\cdot526$ Å. The relative amounts of s- and p-character in an $sp^{4\cdot12}$ orbital are given by the relation per cent $p = 4\cdot12 \times$ per cent s; the sum of the per cent s- and per cent p-character in the hybrid is 100 per cent hence $4\cdot12 \times$ per cent $s +$ per cent $s = 100$ and therefore the per cent $s = 19$ per cent. The two carbon orbitals of cyclopropane carrying hydrogen atoms can be shown similarly to have 31 per cent s-character and 69 per cent p-character; these orbitals are, therefore, $sp^{2\cdot28}$.

The Tricovalent Carbon Atom

The tetravalent tetrahedral carbon atom was the cornerstone of classical theory; the quantum mechanics advanced the classical

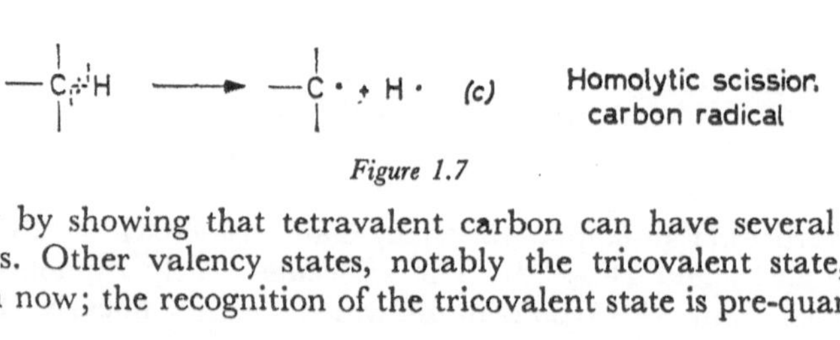

Figure 1.7

theory by showing that tetravalent carbon can have several geometries. Other valency states, notably the tricovalent state, are known now; the recognition of the tricovalent state is pre-quantum theory and came from the results of synthetical organic chemistry. The three important tricovalent states are called the carbonium ion, the carbanion and the carbon radical; the recognition of the first two of these states has been gradual but the trivalent radical was recognized definitely in 1900 (Gomberg[26]). In electronic terms these three states may be imagined to arise from the three different modes of scission possible for a carbon-hydrogen electron pair bond (*Figure 1.7a, b* and *c*).

The carbonium ion (*cationic carbon*)—The heterolytic scission of a C—H bond (*Figure 1.7a*) gives a carbonium ion; the bonding pair remains attached to the hydrogen atom which separates as a hydride anion. The $\overset{+}{C}$ ion may be imagined also to be derived from a $1s^2 \cdot 2s \cdot 2p \cdot 2p \cdot 2p$ carbon atom by the removal of the $2p$ electron since of the electrons present, the $2p$ electrons are ionized most easily because of their higher energy. The resultant $1s^2 \cdot 2s \cdot 2p \cdot 2p \cdot$ state can provide three coplanar sp^2 hybrids which form the three covalencies of the ion; the $\overset{+}{C}$ ion is, therefore, an ionized ethylenic carbon atom. The evidence for the planarity of the $\overset{+}{C}$ is indirect but is supported by the isoelectronic principle, thus $(CH_3)_3\overset{+}{C}$ is electronically identical to $(CH_3)_3B$ and isoelectronic particles are expected to have the same geometry. Physical methods show a planar geometry for $(CH_3)_3B$ with $\overset{\frown}{CBC} = 20$ degrees and this geometry is expected for the $\overset{+}{C}$ ion. A stereochemistry other than the sp^2 planar may be deduced for $\overset{+}{C}$, thus, the removal of an electron from sp^3 carbon gives a tetrahedral $\overset{+}{C}$ ion. A carbonium ion could be derived, in principle, from sp carbon but feeding three electrons into the orbitals of this atom will give only two equivalent bonds and it is a reasonable assumption that the three covalencies of a $\overset{+}{C}$ ion are identical hence the sp model is invalid. Chemical (Bartlett and Knox[27]) and physical (Franklin and Field[28]) evidence shows that of the sp^3 and sp^2 models the latter is the most stable geometrical form for $\overset{+}{C}$ ions.

The carbanion (*anionic carbon*)—This atomic state also arises by C—H heterolysis but in this case the bonding pair remains on the carbon atom and a proton separates (*Figure 1.7b*). The anionic state may be regarded as deriving from a $1s^2 \cdot 2s^2 \cdot 2p \cdot 2p \cdot 2p \cdot$ carbon atom and again various stereochemical possibilities arise. Thus, the five electrons may be distributed in four sp^3 orbitals so that one such hybrid has a pair of electrons. The stereochemistry of $\overset{..}{C}H_3$, on this model, is essentially that of methane since the stereochemical effect of a lone pair of electrons must be about the same as if a C—H bond existed. Alternatively the $\bar{C}$ ion may be planar based on an sp^2 carbon with the pair of electrons occupying the p-orbital. The tetrahedral carbanion appears to be the most stable geometrical form (Cram[29]) but it has not a great stability relative to the planar form. Thus, at low temperatures ($-70°$) the tetrahedral geometry is maintained but at room temperature a dynamic interconversion between tetrahedral and planar forms occurs.

The carbon radical—Homolysis of a C—H bond (*Figure 1.7c*) gives a carbon radical; this radical is derived from $1s^2 . 2s . 2p . 2p . 2p .$ carbon and again either sp^3 or sp^2 orbitals may be used. An sp^3 radical forms three covalencies using sp^3 orbitals and the odd electron uses the remaining sp^3 orbital; the sp^2 radical has three planar covalencies and the odd electron occupies the p orbital; this latter model is identical to an ethylenic carbon atom. The planar sp^2 geometry is accepted generally for the geometry of the ·CH₃ radical but, as with carbanions, it is not a rigid geometry since it differs little in energy from the sp^3 model (Skell, Woodworth and McNamara[30]). At ordinary temperatures, therefore, the ·CH₃ radical is easily interconverted between the planar and tetrahedral geometries.

One concludes that the isolated carbonium ion and the carbon radical may be described as planar whereas the carbanion has a non-planar, tetrahedral geometry. The radical and carbanion have unstable geometries and, under ordinary conditions, are in a state of dynamic interconversion between the tetrahedral and planar states whereas the planar $\overset{+}{C}$ ion is a fairly stable geometry.

Combinations of Carbon Atomic States

The six carbon states described namely sp^3, sp^2, sp, $\overset{+}{C}$, $\bar{C}$ and C· are idealized states and apply only to isolated atoms. However, the described geometries are found essentially in the real systems CH_4, $CH_2=CH_2$, $CH\equiv CH$, $\overset{+}{C}H_3$, $\bar{C}H_3$ and ·CH₃. These systems may be regarded as parents from which all the complex hydrocarbon systems of organic chemistry can be derived. However, the combination of the six fundamental systems is frequently not additive due to interactions between the combined states. These interactions have important stereochemical consequences and the geometry of a complex molecule becomes constitutive. The geometrical effects in hydrocarbon systems derived from combinations of the fundamental states can now be considered.

18

sp^3—sp^3 combinations—The ethane molecule is an example of a combination of sp^3 carbon atoms. The geometry of this molecule, in simple terms, is a combination of two tetrahedral units with all bonds as σ-bonds, however, a complete analysis requires the consideration of a further effect. The ethane carbon atoms use sp^3 orbitals and these have a 75 per cent *p*-character; the specific

II III IV

orientation of the C_1—H and C_2—H bond orbitals as in (II) (eclipsed orientation) permits, by analogy, with ethylene, the lateral overlap of the orbitals because of their extensive *p*-character. This lateral overlap of orbitals is optimum in the orientation (II) and a rotation of parts of the molecule about C_1—C_2 decreases the overlap. The bonds in ethane have been considered previously as of the localized type, however, the possibility of overlap introduces the important matter of delocalized electrons. Thus, the bonding pair in C_1—H, for example, is no longer restricted to bicentric motion but may move over the C_1 and C_2 nuclei as well as the two hydrogen nuclei. This change in the electron distribution due to delocalization will affect the physical properties and some of these effects are of stereochemical importance. Thus, it is easily seen that overlap of orbitals leads to bonding between C_1 and C_2 so that the established single bond acquires some double bond character. The carbon-carbon bond in ethane is, therefore, best described as a partial double bond and rotation about it is less free than for a pure single bond. This type of interaction of atomic states, which is small, is called 'second order hyperconjugation' (Mulliken, Reike and Brown[31]); partial double bond formation decreases the bond length and assuming a C—C bond length of 1·54 Å in ethane it has been calculated that this length would be 1·58 Å in the absence of hyperconjugation. It should be pointed out that the geometry (II) although it is stabilized by hyperconjugation is not necessarily the most stable geometry for the molecule (see page 50) nevertheless the molecule can adopt geometry (II) and hyperconjugation can operate. Alicyclic molecules such as a cyclohexane (III), its higher homologues and polycyclic derivatives are made up of sp^3 carbon atoms. The geometry (III) is the most stable one for cyclohexane and it is to be noted that the orientation of the sp^3 units does not permit second order hyperconjugation.

sp^3—sp^2 combinations—The propylene molecule CH_3—$CH=CH_2$ is an example; the geometry (IV) of this molecule shows the possibility of overlap of the sp^3 orbital of C_1 with the π-orbital of the double bond. The geometry of a π-orbital makes greater overlap possible in propylene than in ethane; this type of interaction, which is still small, is called 'first order hyperconjugation'. The stereochemical consequences are as described for ethane 'namely' hindrance of rotation about the C_1—C_2 bond due to a partial double bond character together with the shortening of this bond.

sp^3—sp combinations—This system is illustrated by methylacetylene CH_3—$C{\equiv}CH$; first order hyperconjugation operates but the electron clouds of the two π-bonds permit greater overlap and the hyperconjugation effect is enhanced relative to propylene. Bond shortening is very small in propylene and has not been detected but in methylacetylene a shortening of 0·08 Å has been established (Badger and Bauer[32]).

sp^3—$\overset{+}{C}$ combinations—The ethyl cation CH_3—$\overset{+}{C}H_2$ contains these states and hyperconjugation occurs similar to that in propylene; the hyperconjugation operates by the use of a methyl group sp^3 orbital and the vacant p orbital of the cation. Hyperconjugation involving ions is a much more important effect than in molecules (Muller and Mulliken[33]). The effect involves partial double bond formation in a classical single bond so increasing the strength of this bond and increasing the stability of the molecule. The hyperconjugative stabilization for propylene is 3 kcal/mole but for the ethyl cation it is 36 kcal/mole.

sp^3—$\overline{C}$ combinations—This type of combination, for example CH_3—$\overline{C}H_2$, simulates ethane stereochemically and because of hyperconjugation. However, hyperconjugation in this type of combination involves the lone pair of the anionic centre and so is a more important effect than in ethane.

sp^3—$C\cdot$ combinations—Hyperconjugation in a system such as the ethyl radical CH_3—$CH\cdot_2$ is similar to that in the carbonium ion $C_2\overset{+}{H}_5$ since an sp^3 and a p orbital are used. The stabilization due to hyperconjugation in the ethyl radical is 5 kcal/mole and thus is greater than in propylene but less than in $C_2\overset{+}{H}_5$. Hyperconjugation

in systems of the last three combinations will be optimum if $\dot{C}$, $\bar{C}$ and C· are planar since overlap with an sp^3 orbital will be maximum; these states therefore, will tend towards the planar or be stabilized as planar in the absence of other factors; the relative importance of the various factors will be discussed later.

sp^2—sp^2 *combinations*—Hyperconjugative interaction is of recent recognition experimentally and theoretically (Baker and Nathan[34], Mulliken, Reike and Brown), however, sp^2—sp^2 type interactions have been known to organic chemists for many years although a detailed interpretation is only possible in quantum mechanical terms. The sp^2—sp^2 combination is, in fact the 'conjugated' system of classical organic chemistry.

The special nature and stability of 'conjugated' systems was recognized in the classical period to the extent that special theories (Thiele[35]) were developed in order to account for them; these theories related to chemical properties but the interactions also have important geometrical effects. The simplest sp^2—sp^2 combination is the molecule of but-1,3-diene, $CH_2=CH-CH=CH_2$; at first sight the molecule is expected to be a simple combination of two π-ethylenic units but delocalization interaction gives a quite different molecular orbital picture.

The previous use of molecular orbital theory has been merely to indicate that orbital overlap provides the condition for molecular orbital formation and that localized or delocalized orbitals may be present in molecules; it is now necessary to indicate briefly some further fundamental ideas of molecular orbital theory. This theory envisages the construction of molecules in terms similar to that for atoms. Thus, for atoms the starting point is the bare nucleus and, by the 'aufbau' principle electrons are fed into the vacant orbitals starting with the one of lowest energy. The introduction of the first electron into the lowest energy orbital is followed by a second electron of opposed spin; this orbital is filled by the two paired electrons (Pauli principle). The complete atom is built by feeding pairs of electrons into orbitals of increasing energy. Molecules are constructed in a similar way; the bare nuclei are considered spatially distributed as in the molecule; pairs of electrons of opposed spin are then fed into molecular orbitals about the nuclear skeleton so building up the molecule. This picture may be simplified by considering the group of nuclei and their appropriate inner shells of electrons as the starting structure; the peripheral valency electrons are then fed into molecular orbitals about this complex 'nucleus'. Molecular orbitals are similar to atomic orbitals in that they have characteristic shapes and energies and are defined by quantum numbers; they differ in that they are polycentric. The feeding of pairs of electrons into orbitals of increasing energy can lead to the completion of molecular shells and sub-shells; completed shells, as for atoms, correspond to very stable electronic groupings and are important in aromatic systems.

A localized molecular orbital may be pictorially related to two overlapping component atomic orbitals, however, the formation of a molecular orbital by overlap is not an indiscriminate process and certain conditions must be fulfilled. Thus, the component atomic orbitals must be of similar energy and they must have the same symmetry relative to the internuclear axis. The $1s$ orbital of a hydrogen atom cannot combine with the $1s$ orbital of a chlorine atom since the more positive nucleus of the latter results in a much lower energy of $1s$ electrons. Two atoms may contain a set of mutually perpendicular p orbitals (along axes x, y and z), the symmetry condition allows molecular orbital formation only by the overlap of p_x and p_x, p_y and p_y, p_z and p_z; the overlap may be linear or lateral. Molecular orbitals have different geometries to the component atomic orbitals; one mathematical procedure relates molecular orbitals to component atomic orbitals in a simple way (the 'linear combination' method) and this procedure shows that two atomic orbitals provide two molecular orbitals; each molecular orbital can hold an electron pair. Thus, the combination of the $1s$ orbitals of two hydrogen atoms gives two molecular orbitals of different energies; the two $1s$ electrons are fed into the lowest energy orbital and, as theory shows, this results in their concentration between the nuclei and a stable molecule is formed. The shape and electron distribution of the higher energy orbital is such that there is a paucity of charge between the nuclei and no hydrogen molecule would result. The lowest energy orbital is called a bonding orbital; the higher energy orbital, which has a node between the nuclei, is called an anti-bonding orbital. Other molecular orbitals can be constructed about two nuclei by combining $2s$, $2p$ etc. atomic orbitals; these orbitals apply only to excited molecules. Two helium atoms have four electrons and in the ground state molecule these electrons are fed in pairs into the two lowest orbitals. The destabilization due to the filled anti-bonding orbital exceeds the stabilization due to the filled bonding orbital so that a stable He_2 molecule is impossible. In the ethylene molecule the π-orbitals are of most interest and attention may be focused on them; the system of nuclei and inner shells and the localized

22

σ-orbitals may be regarded as the 'nucleus'. The carbon p orbitals overlap and a bonding and anti-bonding π-orbital are formed; the π-bond is established by feeding the pair of electrons into the lowest energy bonding orbital.

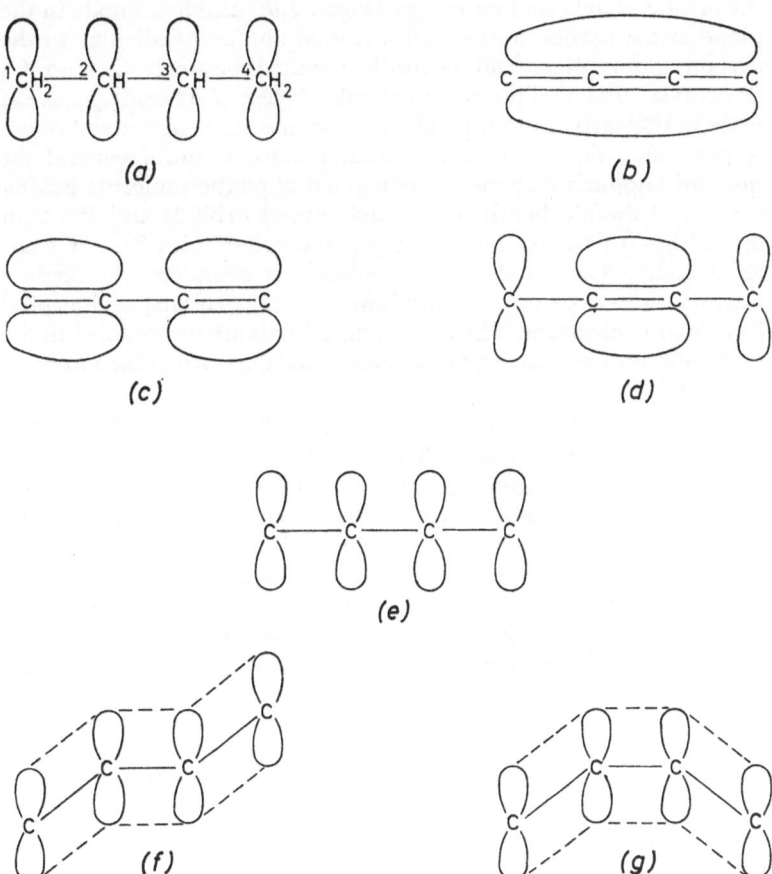

Figure 1.8

Delocalized molecular orbitals follow the same principles as localized orbitals but, of course, they are polycentric not bicentric ones. The sp^2—sp^2 type of combination has delocalized orbitals and the but-1,3-diene molecule is an example. This molecule has four parallel p orbitals, one on each carbon atom, the other bonds present are of the σ-type (*Figure 1.8a*). The geometry of the molecule

c

permits overlap of all the p orbitals and four molecular orbitals obtain. These orbitals are tetracentric, covering all carbon nuclei, and are of different geometry and energy. The orbital of lowest energy (*Figure 1.8b*) has a node in the plane of the carbon atoms; the orbital of next highest energy (*Figure 1.8c*) also has a node in the plane of the carbon atoms and a second one perpendicular to the carbon atom plane and centrally situated between C_2 and C_3 (*trans*-axial node). The other orbitals (*Figures 1.8d* and *e*) have a node in the carbon atom plane and two and three *trans*-axial nodes respectively. This molecular orbital picture is quite general for sp^2—sp^2 aliphatic systems; a conjugated aliphatic molecule having n classical double bonds has $2n$ delocalized orbitals and the $2n$th orbital has a node in the carbon plane together with $2n - 1$ *trans*-axial nodes. The but-1,3-diene molecule is completed by feeding the four p electrons into the two lowest energy orbitals; each orbital has a pair of electrons. The remaining orbitals are unoccupied in the molecular ground state but can become occupied when the molecule is excited.

The above representations are diagrammatic; the molecule of but-1,3-diene actually has $\widehat{CCC} = 120$ degrees. In terms of the symmetry or maximum overlap criterion molecular orbital formation requires the component p orbitals to be parallel; two parallel arrangements are possible and both are planar (*Figure 1.8f* and *g*). If but-1,3-diene was made up of two classical ethylenic units the two arrangements would be interconverted easily by rotation about the single C_2—C_3 bond; molecular orbital theory modifies this picture. Thus, the electron pair in the lowest energy orbital provides a charge cloud of about the same charge density, between all four carbon atoms of the molecule. The electrons in the other orbital (*Figure 1.8c*) contribute a greater charge density between C_1—C_2 and C_3—C_4 than between C_2—C_3 since a node exists between these latter two atoms. The ground state molecule, therefore, has a rather higher density of charge between C_1—C_2 and C_3—C_4 than between C_2—C_3; a description in terms of bonds is that the classical single C_2—C_3 bond actually has some double bond character although this is less than the double bond character of the C_1—C_2 and C_3—C_4 bonds. The bond picture in terms of atomic orbitals may be interpreted as a quite strong overlap of the p-orbitals on C_1—C_2 and C_3—C_4 and a weaker overlap of these on C_2—C_3. The partial double bond between C_2—C_3 is weak but real; an energy of only 5 kcal/mole is required to induce rotation about it. Thus, although the two planar geometries for but-1,3-diene are the most stable, because of optimum parallel orbital overlap, they are easily inter-converted under ordinary conditions. As will be shown later other factors are involved in the stabilities of *Figure 1.8f* and *g* and these favour the geometry of *Figure 1.8f*; experimental evidence shows that at ordinary temperatures the substance is made up largely of molecules of the type of *Figure 1.8f* but at higher temperatures more molecules have the geometry of *Figure 1.8g* (Walsh[36]).

The delocalized orbital picture interprets the planar geometry and restricted rotation of but-1,3-diene but it is also superior for energetic reasons to the alternative localized picture. The localized picture uses separate π-orbitals about C_1—C_2 and C_3—C_4; the electron motion is bicentric and, as the electrostatics show, the more nuclei moved over by electrons the lower their potential energy. Delocalization clearly stabilizes a molecule and the potential energy difference between the localized and delocalized models is a quantity of the greatest chemical importance; it is called the 'delocalization or resonance energy'. The delocalization energy for but-1,3-diene is 3·5 kcal/mole. It will be clear that the geometrical effects of de-localization are (*a*) the enforcement of molecular planarity since planarity is required for optimum overlap and (*b*) the acquisition of partial double bond character by classical single bonds; this effect shortens the classical single bond and makes rotation about it more difficult. The C—C single bond lengths in but-2-diene, CH_3—CH=CH—CH_3, are about 1·54 Å; the C_2—C_3 bond length in but-1,3-diene is 1·47 Å. The geometrical effects in sp^2—sp^2 systems are explicable only in modern terms; the stabilization effects were recognized classically but, of course, could not be interpreted.

Cyclic sp^2—sp^2 systems exist and the benzene molecule is the example 'par excellence'. Kekulé[37] postulated a regular planar hexagonal structure (V) for the molecule and added the double bonds to preserve the tetravalency of carbon. This theory was in fact the first stereochemical theory although, of course, it was of very limited form. The van't Hoff theory predicts a planar model for benzene but this is in error in other respects since the molecule has neither tetrahedral angles nor is it a system of alternating double and single bonds. The geometry of the benzene molecule derived by physical methods (Lonsdale[38]; Karle[39]) is as shown in (VI); it is a planar regular hexagon having C—C = 1·39 Å, C—H = 1·08 Å and $\widehat{CCC}$ = $\widehat{HCC}$ = 120 degrees. The quantum mechanical picture, as anticipated from the geometry of the molecule, invokes sp^2 carbon atoms in order to construct the ring. The carbon atoms

use two sp^2 orbitals to form σ-bonds with each other and these
bonds have $\widehat{CCC} = 120$ degrees. The third sp^2 orbital of the
carbon atom forms a σ-bond with a hydrogen atom by overlap
with a $1s$ orbital so that $\widehat{HCC} = 120$ degrees. Each carbon atom

has a remaining p orbital containing an electron and the six p
orbitals are oriented perpendicular to the plane of the ring as in
(VII). This orientation permits overlap and a group of six molecular
orbitals can result; these orbitals are delocalized and of different
energy in some cases. The lowest energy orbital is similar to that in
but-1,3-diene; it extends round the ring above and below the ring
plane and has a node in the plane of the ring. The next two orbitals
are equal in energy; they have nodes in the ring as indicated by
the broken lines in (VIII) and (IX). The remaining three orbitals
are not involved in the ground state benzene molecule. The six p
electrons enter, as pairs, the three lower energy orbitals so com-
pleting the molecule. A more general discussion of the orbital
picture of benzene is instructive (Hückel). The group of molecular orbitals derivable consist of a lowest energy orbital with one
node, as described, and a series of pairs of orbitals; each pair has
a different energy but the orbitals of the pair have identical energies
(degenerate). The filling of pairs of degenerate levels corresponds
to the completion of a molecular orbital shell; this is analogous
to the completion of an inert gas structure and constitutes a very
stable electronic group. The benzene molecule has a completed
molecular orbital shell and so especially high stability; this constitutes the quantum mechanical interpretation of the long recognized 'aromatic sextet'. The special form of delocalization associated
with shell completion is in complete agreement with the planarity,
bond lengths and stability of the benzene molecule; a high delocalization energy is expected and it has been evaluated as 36

kcal/mole (Pauling and Sherman[40]) though this value may well be too high (Dewar[41]).

The above theoretical description suggests that the benzenoid or aromatic properties of planarity and stability will appear in any cyclic system so constituted that its annular atoms participate in a single conjugated system and if it has pairs of degenerate orbitals filled. The discussion of benzene shows that orbital pairs will be filled if the molecule has $4n + 2(n = 1, 2, \ldots)$ conjugation electrons (Hückel's rule). The unknown cyclobut-1,3-diene (X) is the simplest conjugated cycle; the molecule is expected to be planar and to have some stability but this cannot be very great since the molecule has only four conjugated electrons and does not obey Hückel's rule. The related molecule (XI) is established as planar but again it is not aromatic in stability. The cyclopentadiene (XII) and cyclo-heptatriene (XIII) are not aromatic since annular conjugation is broken. The molecule (XIV) should be aromatic but apparently other factors intrude to affect the planarity and hence the conjugation. The molecule (XV) obeys Hückel's rule and has aromatic properties but (XVI) does not obey the rule and is not aromatic. The cyclo-octatetrene molecule (XVII) is expected to be planar but not highly stable since it has only eight electrons. Planar cyclo-octatetrene has $\widehat{CCC} = 135$ degrees and this requires a change from sp^2 carbon; the destabilization induced is not compensated for by the conjugation resulting from planarity so that the molecule is a collection of isolated ethylenic units and has the puckered 'tub' form (XVIII) (Lippincott and Lord[42], Hedberg[43]). Azulene (XIX) and tropolone (XX) are both aromatic; an aromatic sextet of electrons in tropolone arises from the three double bonds, the π-electrons of the $C = O$ reside on the oxygen atom leaving a vacant carbon orbital to participate in conjugation (XXI). The conjugation results in the distribution of positive charge and tropolone is best represented as (XXII); the internal circle is drawn to represent annular conjugation, alternatively a broken line, as in (XXIII) is used. The parent molecule of this aromatic series is tropone (XXIV) and clearly tropolone is related to (XXIV) as phenol is to benzene. The polycyclic molecules naphthalene (XXV), phenanthrene (XXVI) and diphenyl (XXVII) are typically aromatic species and are planar and stable. The resonance energies follow from the degree of the delocalization; phenanthrene 130 kcal/mole, diphenyl 91, naphthalene 77, azulene 45·5, tropolone 28·6 and cycloheptatriene 6·7 kcal/mole. One concludes that in conjugated systems electron delocalization results in molecular stability, planarity and the

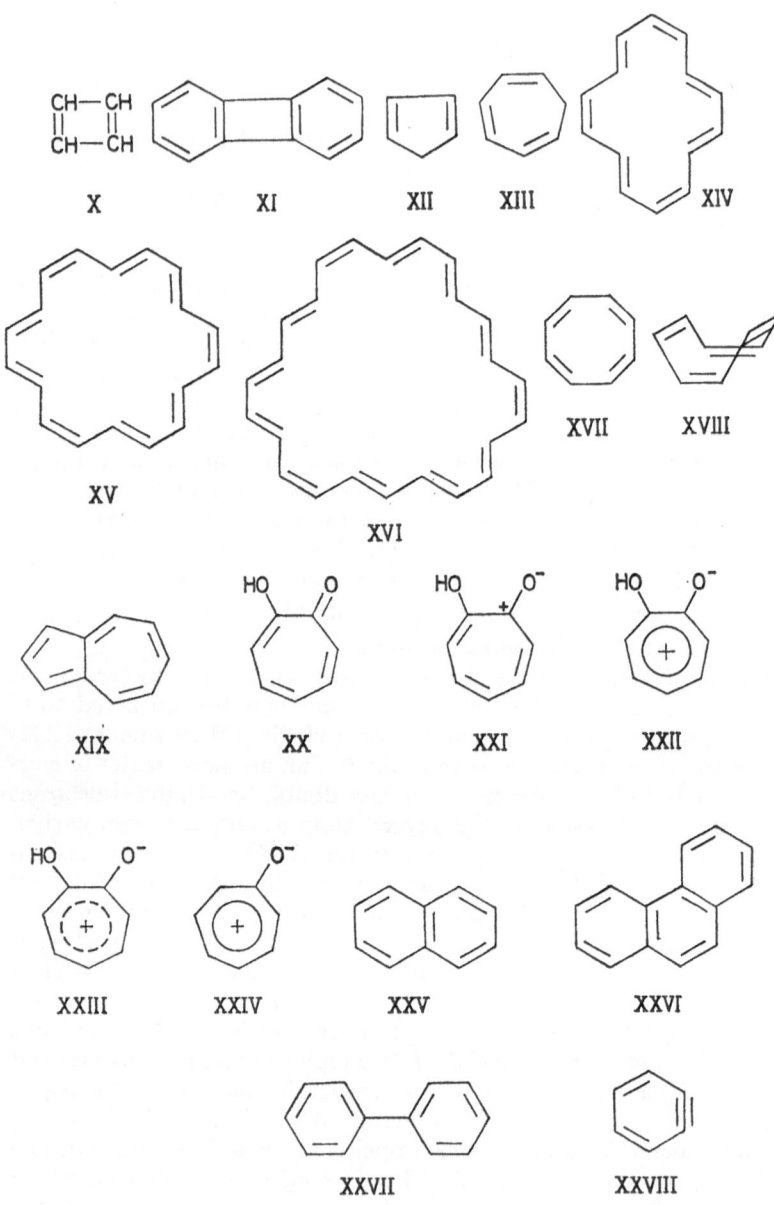

modification of bond lengths; these effects are manifest in open or cyclic systems provided no other factors operate to inhibit or prevent conjugation; in cyclic systems following Hückel's rule these effects manifest themselves to an extreme degree and the properties are called aromatic, thus aromaticity results in rigid planarity, high stability and equal bond lengths. While the properties of conjugation and the possession of $(4n + 2)$ electrons usually lead to aromaticity it appears that large ring compounds for example, (XIV) may be exceptions (Longuet-Higgins[44], Coulson[45]).

sp^2—sp combinations—Aliphatic molecules of this type, for example, $CH_2=CH—C\equiv CH$ are entirely analogous to the sp^2—sp^2 type discussed above. A cyclic sp^2—sp structure of recent interest is the reactive 'intermediate' benzyne (XXVIII); this molecule cannot be constructed from a ground state linear acetylenic unit and an excited unit is used. This excited state is a bent structure and allows the formation of a cycle.

sp^2—$\overset{+}{C}$ combinations—The allyl carbonium ion $CH_2=CH—\overset{+}{C}$ is an example; in this system the p orbitals of the ethylenic system can overlap with the vacant p orbital of $\overset{+}{C}$. The conjugation results in such systems developing the characteristics of stability, planarity and modified bond lengths. Thus, the $C—\overset{+}{C}$ bond acquires partial double bond character, it is shortened and rotation about it is hindered.

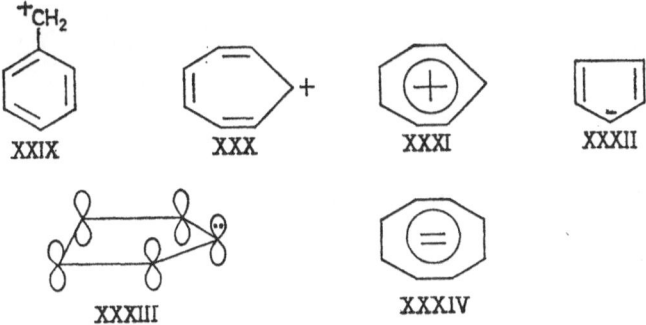

The benzyl carbonium ion (XXIX) has increased overlap possibilities and has greater stability. Carbonium ions are usually transient intermediates in chemical reactions, however, certain $\overset{+}{C}$ ions are highly stable as a consequence of conjugation. Thus, the cycloheptatrienylium ion (XXX) (tropylium or tropenium ion) has its $\overset{+}{C}$ as part of a conjugated system having a sextet of electrons (Doering and Knox[46]; Hückel). This ion, better represented as (**XXXI**) is therefore, aromatic hence its high stability.

sp^2—$\overline{C}$ combinations—The isolated carbanion has been described as tetrahedral sp^3 but easily capable of attaining the planar sp^2 state. The cyclopentadientyl anion (**XXXII**) will have optimum overlap of orbitals and aromatic (sextet) properties if the $\overline{C}$ has an sp^2 structure with the pair of electrons occupying the perpendicular p orbital as in (**XXXIII**). The high stability of the planar aromatic structure for the ion therefore enforces sp^2 hybridization for the carbanion in this molecule. The dianion (**XXXIV**) is derived by the abstraction of two electrons from cyclo-octatetrene leaving a sextet of electrons; (**XXXIV**) is therefore aromatic.

sp^2—$C\cdot$ combinations—The allyl and benzyl radicals are examples; the radical carbon atom is planar and optimum overlap of p orbitals occurs resulting in stability and planarity.

sp—sp, sp—$\overset{+}{C}$, sp—$\overline{C}$ and sp—$C\cdot$ combinations—These combinations of units are analogous to those described above and the same stereochemical results obtain.

sp^2—sp—sp^2 combinations—The symbolism refers to a molecule such as allene $CH_2=C=CH_2$. The stereochemistry of the allenes was predicted by van't Hoff (1875) using tetrahedral carbon atoms; this classical model (**XXXV**) shows that the hydrogen atoms lie

XXXV   XXXVI

in perpendicular planes. The quantum mechanical model has a similar stereochemistry but shows that C_1 and C_3 are sp^2 atoms and that C_2 is an sp atom as shown in (**XXXVI**).

THE NITROGEN ATOM

Historically the hypothesis of the tetrahedral carbon atom arose when certain problems, requiring a geometrical solution, were encountered. No analogous problems existed at the time for nitrogen containing compounds but the success with carbon did lead to thinking about the stereochemistry of other atoms. An initial difficulty with the stereochemistry of nitrogen was that the classical valency theory allowed both trivalent and pentavalent states for this atom. The development of classical electronic theory (1916) and the accumulation of experimental evidence (e.g. Schenk[47]) clearly established that the nitrogen atom can be tricovalent and tetracovalent at the most; these states are understood now in quantum mechanical terms.

The Tricovalent Nitrogen Atom

The quantum mechanics shows that the tricovalent nitrogen atom can have various geometries; the tricovalent state is more common in organic molecules than is the tetracovalent state and so will be discussed first.

The p^3 or pyramidal nitrogen atom—The ground state nitrogen atom is $1s^2 . 2s^2 . 2p . 2p . 2p$; it is isoelectronic with the carbanion. The three unpaired $2p$ electrons occupy three mutually perpendicular orbitals and so can form three covalencies along these directions. It was thought earlier that this description sufficed to define the bond type in tricovalent species such as ammonia and aliphatic amines. However, the description is incomplete since the $H\widehat{N}H$ in NH_3 is 107 degrees and the $C\widehat{N}H$ in CH_3NH_2 is 108 degrees; other factors must operate to widen the 90 degree angle. The angle widening may be accounted for by assuming partial hybridization by some admixture of the $2s$ orbital with the $2p$ orbitals. If the three orbitals of nitrogen in NH_3 which bond with hydrogen are 90 per cent p and 10 per cent s then the remaining orbital is 30 per cent p and 70 per cent s; the lone pair of electrons occupies this fourth orbital.

The sp^3 or tetrahedral nitrogen atom—The partial sp^3 hybridization found in aliphatic amines is complete, apparently, in aromatic amines such as aniline $C_6H_5NH_2$ since the $H\widehat{N}C$ is 110 degrees (Wepster[48]). Tetrahedral tricovalent nitrogen is analogous to the carbanion; three sp^3 hybrids contain an electron each and the fourth

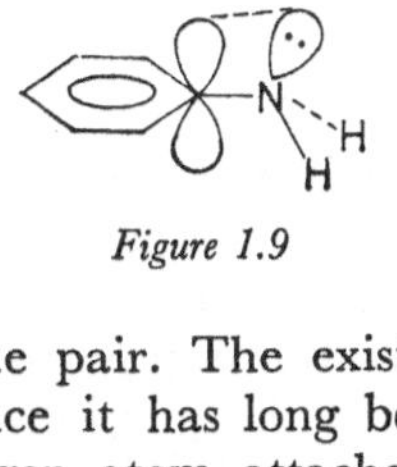

Figure 1.9

orbital contains the lone pair. The existence of sp^3 nitrogen in aniline is surprising since it has long been known (Robinson[49]) that a tricovalent nitrogen atom attached to an sp^2 unit, as in vinylamine, or to an aromatic ring, as in aniline, is capable of participating in conjugation by using the lone pair. The situation in molecular orbital terms is described as an overlap of carbon π-orbitals and the lone pair orbital of the nitrogen atom. Thus, it is expected, for energy reasons, that an sp^3 nitrogen atom in aniline would become an sp^2 state as is found for a conjugated carbanion. However, in spite of the reduced overlap the sp^3 state is retained in aniline and it is an sp^3 lone pair orbital which overlaps with the π-orbitals of the ring (*Figure 1.9*).

The sp^2 or planar nitrogen atom—The planar sp^2 nitrogen atom, which is found in the azo unit —N=N—, is not entirely analogous to ethylenic carbon in electron distribution because carbon is tetracovalent. The nitrogen atoms in azobenzene each use an sp^2 orbital for σ-bonding with each other and a second sp^2 orbital for bonding with an aromatic ring. The nitrogen p orbitals π-bond with each other and the lone pair occupies the remaining sp^2 hybrid (*Figure 1.10a*). An sp^2 nitrogen atom, together with an sp^2 carbon atom, makes up the azomethine, —CH=N—, unit. Amido nitrogen is planar and sp^2 but is different from both azo and azomethine nitrogen in electron distribution. Thus, the amido nitrogen uses the three sp^2 hybrids to form three covalencies; the electron pair occupies the p orbital (*Figure 1.10b*). One concludes that an

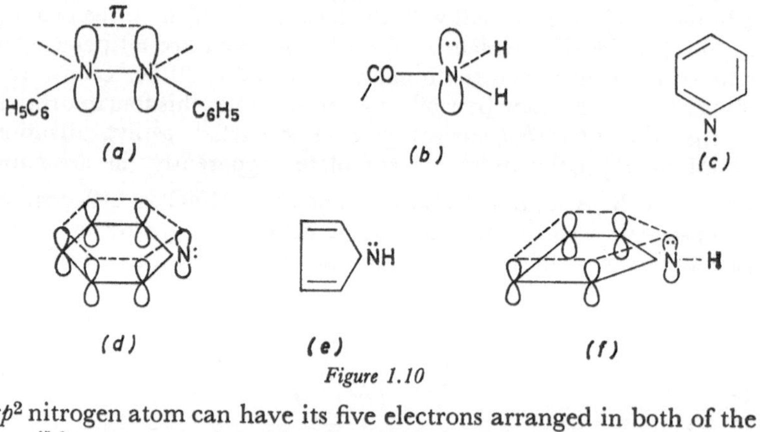

Figure 1.10

sp^2 nitrogen atom can have its five electrons arranged in both of the possible ways namely with a lone pair in either an sp^2 orbital or in the p orbital. These two arrangements are illustrated further by the heterocyclic molecules pyridine and pyrrole. The pyridine molecule has an azo nitrogen atom; the p orbital with its electron forms part of the conjugated system of six electrons (*Figure 1.10c and d*);

the molecule is clearly analogous to benzene and is aromatic. The nitrogen atom in pyrrole is the amido type; the p orbital has a lone pair and by overlap these conjugate with the carbon electrons again giving a sextet and so an aromatic molecule (*Figure 1.10e* and *f*); pyrrole is clearly the nitrogen analogue of the cyclopentadienyl anion.

The sp or linear nitrogen atom—The cyano group contains sp nitrogen; an sp orbital of the carbon forms a σ-bond with an sp nitrogen orbital. The p orbitals of nitrogen form two π-bonds with two p orbitals of carbon and the nitrogen lone pair resides in the other sp orbital.

The Tetracovalent Nitrogen Atom

The five valency electrons of the $1s^2 . 2s^2 . 2p . 2p . 2p$. nitrogen atom occupy the orbitals of the $2s$ and $2p$ sub-shells; the orbitals which can be used for bond formation are restricted to these four (or hybrids of them) since although five electrons are available five covalencies can be formed only by promoting a $2s$ electron to the $3s$ level; this process requires too much energy. Thus the nitrogen atom has a maximum covalency of four and of these valencies three are ordinary covalent bonds while the fourth is a co-ordinate covalency formed by the nitrogen atom sharing its lone pair with an electron acceptor.

The sp³ or tetrahedral nitrogen atom—The detection and interpretation of the various geometries of the tricovalent nitrogen atom required the development of physical methods and of quantum mechanics. Earlier chemical evidence (e.g. Mills[50]) suggested the non-planarity of the valencies in tricovalent nitrogen compounds but, of course, it was not suspected that such atoms would prove to have a variety of geometries. The tetracovalent nitrogen is tetrahedral; this geometry was postulated by Pope and Peachey[51] and by Mills and Warren[52] when it was shown that a square planar or a pyramidal geometry did not accord with the chemical facts. The tetrahedral geometry has been substantiated since by physical methods.

The tetracovalent nitrogen atom has three sp^3 orbitals with one electron and these form ordinary covalencies with other atoms; the lone pair occupies the remaining sp^3 orbital and forms a co-ordinate link with an electron acceptor and the nitrogen atom acquires a positive charge. The sp^3 state is found in quaternary ammonium salts $R_4\overset{+}{N}\overline{C}l$, amine oxides $R_3\overset{+}{N} \to \overline{O}$ and complexes such as $R_3\overset{+}{N} \to \overline{B}F_3$.

The Oxygen Atom

Detailed stereochemical knowledge of the oxygen atom is recent; the atom is capable of a multiplicity of states each with its characteristic geometry. The ground state atom is $1s^2 \cdot 2s^2 \cdot 2p^2 \cdot 2p \cdot 2p$. the six valency electrons must occupy the four orbitals of the L shell (or hybrids of them) and therefore as for nitrogen, the atom has a maximum covalency of four. Tetracovalent oxygen is rare and is not found in simple organic systems but the di- and tricovalent states are well known.

The Dicovalent Oxygen Atom

This is, of course, the classical valency state of oxygen but quantum mechanics shows that it can have several geometries.

The p or angular oxygen atom—The $1s^2 \cdot 2s^2 \cdot 2p^2 \cdot 2p \cdot 2p$. oxygen atom can form two covalent bonds by using its $2p$ electrons; such bonds would be at 90 degrees. The use of essentially pure p orbitals was invoked earlier in theoretical chemistry to interpret the geometry of water and alcohol molecules but the oxygen bond angle is 104 degrees in water and the angle widening is interpreted now in terms of hybrid orbitals. Thus the widening is explained as due to a trend of the bonding orbitals to sp^3 hybrids by some admixture of $2s$ character with the two $2p$ orbitals. The problem is similar to that of ammonia except that the lone pairs of the oxygen atom occupy two equivalent orbitals derived from the third $2p$ orbital, the residual $2s$ orbital and the excess of the p-bonding orbitals.

On correlation theory the water molecule derives from an O^{2-} ion which has four pairs of tetrahedrally distributed electrons. When two protons are brought up to bond the bonding pairs are pulled further from the nucleus and since in addition they have become bonding pairs the tetrahedral angle will decrease. The lone pairs interact less with the bonding pairs but more with each other, the repulsions increase the lone pair angle to greater than the tetrahedral value. The ethylene oxide molecule (**XXXVIa**) probably uses

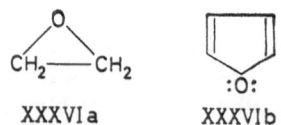

XXXVIa   XXXVIb

pure $2p$ orbitals; the $\widehat{COC} = 61\cdot6$ degrees and so the p orbitals form 'bent' bonds, not σ-bonds, with the carbon atoms.

The sp^3 or tetrahedral oxygen atom—The trend to sp^3 hybridization in H_2O is nearly complete in aliphatic ethers and in the furan molecule (XXXVIb); the $\widehat{COC}$ angle is 108 degrees in these molecules. Two sp^3 orbitals in these molecules are used for covalency formation and the remaining two orbitals carry the lone pairs. An sp^3 lone pair in furan can conjugate with ring electrons as described for aniline and in this respect furan more closely resembles aniline than pyrrole. The conjugated system of furan has a sextet of electrons and is aromatic.

The sp^2 or planar oxygen atom—The carbonyl group, $>C=O$, is made up of an sp^2 oxygen and carbon atom. An sp^2 orbital from the carbon and oxygen atom overlap to establish a C—O σ-bond. The carbon and oxygen p orbitals overlap and establish a π-bond; the two lone pairs occupy the two remaining sp^2 orbitals of the oxygen atom. Phenol and phenol ether molecules also have an oxygen bond angle of 120 degrees showing that sp^2 hybrids are used. In these molecules two sp^2 orbitals are used to establish covalencies and the lone pairs occupy the remaining sp^2 and p orbitals. These molecules are clearly stereochemically similar to pyrrole and the p lone pair is orientated optimally for conjugation with the ring.

The Tricovalent Oxygen Atom

This state is rare; it is exemplified by $H_3\overset{+}{O}$, $R_3\overset{+}{O}\overline{F}$ and $R_2\overset{+}{O} \rightarrow \overline{B}F_3$. The tricovalent state must involve a lone pair and, therefore, a co-ordinate link.

The sp^3 or tetrahedral oxygen atom—Little is known of the geometry of tricovalent oxygen but the hydroxonium ion, $H_3\overset{+}{O}$, appears to be tetrahedral (Ferriso and Hornig[53]).

THE SULPHUR ATOM

The ground state sulphur atom is $1s^2 . 2s^2 . 2p^6 . 3s^2 . 3p^2 . 3p . 3p$; the valency shell contains six electrons, like the oxygen atom, but differs in the important respect that $3d$ orbitals are available for bond formation. Thus, $3s$ and $3p$ electrons can be promoted easily to the $3d$ sub-shell and so all six electrons can be used to form covalencies. A hexacovalent sulphur atom has a dodecet of electrons in the M shell.

The Dicovalent Sulphur Atom

The H_2S molecule has a valency angle of 92 degrees and might appear to use pure sulphur p orbitals. However, it is considered that there is some slight admixture of the $3s$ orbital and this gives slight angle widening. A more complex mixing process, involving a $3d$

orbital, has been proposed (Burrus and Gordy[54]) and this would account for an angle near 90 degrees since admixture of equal amounts of $3s$ and $3d$ orbitals has a cancelling effect on angle widening. An angle of 99 degrees is found in CH_3SH and undoubtedly $3s$ admixture occurs and the bonds therefore tend towards the sp^3 type. The trend is complete, apparently, in thioethers, RSR, and in thiophenols, C_6H_5SH, the angle in these molecules is about 109 degrees and rarely exceeds this value; thiophenol, unlike phenol, does not use sp^2 bonds.

As emphasized previously the use of particular hybrid orbitals by an atom occurs because such orbitals can overlap best with other atoms and so form molecules of optimum stability. This principle is well illustrated by the sulphur atom. Thus, in H_2S the overlap of sulphur p orbitals, having some $3s$ character, with the hydrogen $1s$ orbitals is adequate. However, in an alkyl derivative, such as $(CH_3)_2S$, the sulphur orbitals overlap with sp^3 carbon orbitals; the latter are broader than the $1s$ hydrogen orbitals and so for optimum overlap by sulphur its orbitals acquire more sp^3 character.

The Tricovalent Sulphur Atom

The sulphonium salts exemplify tricovalent sulphur and this group of compounds was of classical interest. Classical valency

XXXVII   XXXVIII   XXXIX

theory represented sulphonium salts (XXXVII) as having tetravalent sulphur and early chemical evidence (Pope and Peachey) suggested that the sulphur atom in these compounds was tetrahedral. Since the classical electronic theory showed that sulphonium sulphur, is tricovalent and that the bromine of (XXXVII) is attached by an ionic bond; sulphonium compounds are represented, therefore, by (XXXVIII). The physical evidence agrees with the expected tetrahedral structure; sp^3 hybrids are used and two of them form ordinary covalencies, a lone pair in an sp^3 orbital forms a co-ordinate link with an electron acceptor (e.g. an R+ in XXXVIII) and the other lone pair occupies the remaining tetrahedral orbital as in (XXXIX). This modern model accounts for the chemical facts just as well as the classical tetravalent tetrahedral model.

The Tetracovalent Sulphur Atom

The sulphoxides and sulphinic esters are now considered to contain tetracovalent sulphur atoms. These molecules were of early interest and on classical valency theory were represented as (XL) and (XLI). The chemical evidence (Harrison, Kenyon and Phillips[55]) refuted the classical structures (XL) and (XLI) and this

$$
\begin{array}{cccc}
\underset{R}{\overset{R}{>}}S{=}O & \underset{R \cdot O}{\overset{R}{>}}S{=}O & R{-}\overset{\cdot\cdot}{\underset{R}{S}}{-}O & R{-}\overset{R}{\underset{O}{S}}{-}O \\
\textbf{XL} & \textbf{XLI} & \textbf{XLII} & \textbf{XLIII}
\end{array}
$$

was supported by electronic theory which allowed only tricovalent sulphur involving a co-ordinate link. The sulphoxide molecule, therefore, was represented with the tetrahedral geometry (XLII). The present theory regards the sulphur-oxygen bond in sulphoxides (and other sulphur-oxygen compounds) as a double bond (Phillips, Hunter and Sutton[56]) and this is electronically interpreted as due to the promotion of a $3p$ electron to a $3d$ orbital. The $3s$ and $3p$ orbitals form essentially sp^3 hybrids and three of these are used for ordinary covalencies, the lone pair occupies the fourth hybrid orbital. The $3d$ orbital overlaps laterally with an oxygen $2p$ orbital forming a π-bond. The geometry of this model clearly is essentially the same as that proposed earlier but the sulphur atom is tetracovalent. The structure and geometry of the sulphinic esters follows analogously.

The Hexacovalent Sulphur Atom

A hexacovalent sulphur is found in sulphones (XLIII), there are four σ-bonds and two π-bonds. The structure and geometry follows from the discussion of sulphoxides but the degree of orbital mixing is more complex as shown by the valency angles in $(CH_3)_2SO_2$; $\widehat{CSO} = 105$ degrees, $\widehat{CSC} = 115$ degrees and $\widehat{OSO} = 125$ degrees. The general stereochemistry of sulphur may be exemplified also by the application of correlation principles. Thus the sulphur atom has six valency electrons, the addition of two more gives S^{2-} as the starting octet species. The eight electrons of this ion are disposed in pairs tetrahedrally and in H_2S two of these pairs bond with protons. These bonding pairs are pulled away considerably from sulphur and so repel less and the angle can close. Angle closing is aided also by lone pair-bonding pair repulsion, these factors account for the diminution of the tetrahedral angle to a value of about 90 degrees. Sulphonium salts have three σ-bonds and a lone pair, the angle will be decreased below tetrahedral. In detail these compounds are regarded as deriving by reaction of S^{2-} with three R^+ and the sulphur will have a net positive charge. Sulphoxides also have three σ-bonds and a lone pair, two R groups bond as R^+ and a pair of electrons of sulphur co-ordinate with an oxygen atom. However, a pair of p oxygen electrons co-ordinate with a d orbital of sulphur establishing a π-bond and a neutral molecule results. Sulphoxides will have bond angles slightly less than tetrahedral; $(CH_3)_2SO$ has $\widehat{CSO} = 107$ degrees and $\widehat{CSC} = 100$ degrees. The sulphones have four σ-bonds and all four pairs of tetrahedrally disposed electrons are used. The hexacovalency of the sulphur atom is completed by the formation of two π-bonds with the oxygen atoms, sulphones will have an irregular tetrahedral geometry as a consequence of the presence of different substituents. The analysis may be carried a stage further, thus, double bonds correspond to a greater concentration of electrons and the repulsion between double bonds is greater than between a double and a single bond and in turn this exceeds the repulsion between two single bonds. These facts account for the $\widehat{OSO} > \widehat{CSC}$ in dimethyl sulphone given above.

The Phosphorus Atom

The configuration of this atom is $1s^2 . 2s^2 . 2p^6 . 3s^2 . 3p . 3p . 3p$; the valency shell has five electrons and these are distributed in a manner similar to those in nitrogen. However, the phosphorus atom is similar to sulphur in having $3d$ orbitals available for use and so can form a maximum of five covalencies.

The Tricovalent Phosphorus Atom

This valency state is analogous to that of tricovalent nitrogen; the $\widehat{CPC}$ in $(CH_3)_3P$ is rather greater than 90 degrees and probably about 100 degrees, thus the $3p$ orbitals have a considerable admixture of $3s$ character and the trend is to sp^3.

The Tetracovalent Phosphorus Atom

Phosphonium compounds contain tetracovalent phosphorus; sp^3 hybrids are probably used and one of these contains a lone pair which co-ordinates with an electron acceptor, for example R^+ in R_4P^+. Chemical evidence which fits the tetrahedral phosphorus atom in phosphonium compounds has been obtained only recently (Holliman and Mann[56a]).

The Pentacovalent Phosphorus Atom

Chemical evidence and the electronic theory shows phosphine oxides to be (XLIV) (Meisenheimer[57]).

XLIV

Modern theory shows that these molecules are similar to sulphoxides in that the P—O bond is double. The promotion of a $3s$ electron to the $3d$ sub-shell is involved and this electron forms a π-bond with oxygen; the other bonds are σ-bonds and are derived from sp^3 hybrids.

THE HALOGEN ATOM

The electronic configurations of the halogen atoms are, F, $1s^2 . 2s^2 . 2p^2 . 2p^2 . 2p$; Cl, $1s^2 . 2s^2 . 2p^6 . 3s^2 . 3p^2 . 3p^2 . 3p$.; Br, $1s^2 . 2s^2 . 2p^6 . 3d^{10}, 4s^2 . 4p^2 . 4p^2 . 4p$; I, $1s^2 . 2s^2 . 2p^6 . 3s^2 . 3p^6 . 3d^{10}, 4s^2 . 4p^6 . 4d^{10} . 5s^2 . 5p^2 . 5p^2 . 5p$. The outer shells of these atoms are very similar and involve s and p electrons. In agreement with classical theory halogen atoms in organic compounds are usually monocovalent but higher valency states are known in compounds such as iodoxybenzene. The monocovalency of halogen atoms derives from their use of their peripheral p electron. Halogen atoms like those of N, O, S and P have lone pairs and in an appropriate molecule these can participate in conjugation. Thus, when a halogen atom is attached to an sp^2 or aromatic carbon atom it forms a σ-bond with its single p electron and by proper orientation a lone pair may overlap with a carbon π-orbital so establishing $p\pi$-conjugation. This effect results in the carbon-halogen bond acquiring some double bond character and being shortened.

THE RESONANCE FACTOR

The previous discussion of molecular structure and geometry has used the molecular orbital theory (Hund[58], Lennard-Jones[59], Mulliken[60]) in a simple qualitative form. An alternative description of molecular structure called the 'valency bond' or 'resonance'

39

theory will now be discussed; this theory, in fact, chronologically preceded the molecular orbital theory.

It was realized even prior to the development of resonance theory that the pure bonds (covalent and ionic) of classical electronic theory rarely existed as such in molecules (Fajans[61]; Sidgwick[62]). In organic chemistry the benzene problem (stability and substitutional reactivity) had existed from Kekulé's time and similar difficulties existed in the interpretation of the chemistry of conjugated systems. Early theoretical attempts were made to deal with conjugated systems (Arndt[63]; Robinson[64]; Ingold[65]) but these were elementary electronic interpretations of the facts of synthetical organic chemistry. Thus, it was recognized that the chemistry of a system such as $CH_2{=}CH{-}CH{=}O$ is not represented adequately by the classical valency picture but if other structures, for example $\overset{+}{C}H_2{-}CH{=}CH{-}\overset{-}{O}$ are invoked then the actual structure and behaviour of the molecule seems to be somewhere between these two extremes, i.e. both of the carbon-carbon bonds exhibit double bond character and the molecule has polarity. The term 'meso-merism' (between the parts) was introduced by Ingold[66] to describe this structure of conjugated systems. The theory of resonance (Pauling) is a generalization and quantum mechanical interpretation of the idea of mesomerism that the actual state of a molecule is not represented by a single classical structure but may be regarded as deriving from the contributions of several valency structures.

The resonance theory starts, of course, from the quantum mechanical atom and the principle of bond formation by the overlap of atomic orbitals; it differs from molecular orbital theory in that localized orbitals only are considered, that is, pairs of atomic orbitals are used. The theory has the great value that because specific pairings are considered the pairing scheme can be represented pictorially by the usual classical line bond method. In considering the pairing all possible pairing schemes are invoked and the actual structure of the molecule is regarded as a hybrid of these. Thus, for the benzene molecule some of the pairing schemes for the p orbitals are shown in *Figure 1.11a, b, c, d* and *e*, these schemes correspond to the classical Kekulé structures (*a* and *b*) and to the Dewar structures (*c, d* and *e*). The resonance theory goes on to say that none of the pairing schemes (structures) represent the actual state of the benzene molecule but that they are intellectual devices the imagined superposition of which gives a picture of the actual molecular structure. Thus, the superposition of the pairing schemes ('canonical states') gives the actual state of the molecule or the 'resonance hybrid'. The contributions of the canonical states to the resonance hybrid need not be identical; the theory shows that the contribution is related to the stability and the greater the stability of a canonical state the greater is its contribution to the hybrid. The Kekulé pairing schemes (*Figure 1.11a* and *b*) for

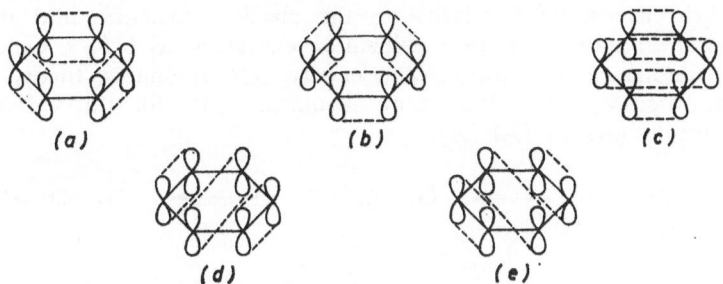

Figure 1.11

benzene have the same stability and are more stable than the Dewar structures. The theory has provided a set of rules which enables the relative stabilities of the various canonical states to be determined; these rules will be illustrated subsequently for particular molecules. The construction of a hybrid molecule for benzene by the superposition of the canonical states gives a molecule having an identical distribution of charge cloud between the carbon nuclei and not a system of alternating single and double bonds; the resonance theory clearly provides the same qualitative picture of benzene as does the molecular orbital theory.

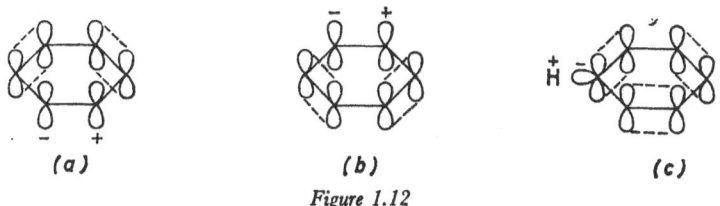

Figure 1.12

A complete description of hybrids also requires the consideration of ionic pairing schemes; these schemes allow canonical states involving σ-bonds to be introduced. Some ionic pairing schemes for benzene are shown in *Figures 1.12a, b,* and *c*; the structures (*Figure 1.12a* and *b*) have two *p* electrons in a *p* orbital so resulting in the development of charge; the structure (*Figure 1.12c*) has the *p*

41

electrons paired in a Kekulé scheme but the pair of electrons of the C—H bond reside in the carbon sp^2 orbital giving an ionic structure. The contribution of ionic structures to essentially covalent molecules is small. The resonance theory, therefore, arrives at a delocalized picture of a molecule by the intellectual device of super-posing a selection of localized structures; the theory shows that the hybrid is stabilized relative to the classical structure and the canonical states and the stabilization energy is, as before, called the resonance or delocalization energy. The resonance theory is illustrated by some important canonical states for the various molecules (XLV)–(XLIX).

$$CH_2\!=\!CH\!-\!Cl \quad\longleftrightarrow\quad \overset{+}{C}H_2\!-\!\overset{-}{C}H\!-\!Cl \quad\longleftrightarrow\quad \overset{-}{C}H_2\!-\!CH\!=\!\overset{+}{C}l$$
XLV

$$CH_2\!=\!CH\!-\!CH\!=\!O \longleftrightarrow \overset{+}{C}H_2\!-\!CH\!=\!CH\!-\!\overset{-}{O} \longleftrightarrow \overset{+}{C}H_2\!-\!\overset{-}{C}H\!-\!\overset{+}{C}H\!-\!\overset{-}{O}$$
XLVI

XLVII

$$CH_3\!-\!I \quad\longleftrightarrow\quad \overset{+}{C}H_3 \quad \overset{-}{I}$$
XLVIII

XLIX

The resonance theory, like the molecular orbital theory, has stereochemical requirements and consequences. The benzaldehyde molecule (XLIX) can have pairing schemes involving the aldehyde group and the ring only if the atomic orbitals are oriented for overlap (*Figure 1.13a*); an aldehyde group not in the plane of the ring (*Figure 1.13b*) cannot interact with the aromatic π-orbitals. Resonance stabilizes a molecule and so groups such as the aldehyde group will take up a specific orientation (*Figure 1.13a*) when attached to an aromatic or similar system. The preferred orientation of groups, due to resonance, results in classical single bonds being shortened due to partial double bond character and rotation about such bonds will be hindered. These stereochemical results are

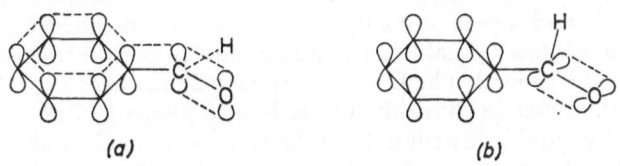

(a)   (b)

Figure 1.13

qualitatively identical to those derived on the molecular orbital theory; both theories will be employed in subsequent discussions.

THE SUBSTITUENT FACTOR

Substituents can affect the geometry of a parent molecule in various ways; here one is concerned with the effect of a substituent on the nature of the orbitals used in the parent molecule. Some examples of the substituent effect have been encountered previously and the effects are of two types namely (a) effects resulting from the interaction of the substituent and the parent molecule and (b) effects occurring without interaction. This first type of effect has been discussed fully in molecular orbital and resonance terms, thus the nitro group in nitrobenzene interacts with the ring and some double bond character of the C—N bond results as well as other geometrical effects. The second type of effect is usually small but it is very prevalent; the introduction of substituents on to sp^3, sp^2 or sp centres slightly affects the geometry of these systems. Thus, in CH_4, sp^3 carbon orbitals are used and this is true also of CCl_4, however in CH_3Br the $\widehat{HCH} = 110$ degrees 48 minutes and this angle opening means that the molecule has three equivalent C—H orbitals, having more s character than an sp^3 carbon, together with a fourth C—Br hybrid having more p character than an sp^3 carbon. The molecule is not regular tetrahedral in agreement with the early belief of van't Hoff and Le Bel. These geometrical effects are not understood fully but clearly are expected in view of the different nature of the substituent; different substituents will affect electronic motion differently and so affect the nature of the orbitals. However, the effects must fundamentally relate to stability as is clearly illustrated by the series H_2S, CH_3SH and $(CH_3)_2S$. As pointed out these substituent effects are generally small; an exception is the 1,1-difluoroethylene molecule. This molecule has the $\widehat{FCF} = \widehat{HCH} = 110$ degrees; the molecule retains the usual π-bond but clearly has carbon atoms using essentially sp^3 orbitals to form the C—F and C—H σ-bonds. The carbon atoms have a $2s$ and three $2p$ orbitals available for bonding, one $2p$ orbital from each is used for the π-bond. The C—F bonds have 25 per cent s and 75 per cent p-character (as have the C—H bonds) hence $\frac{1}{2}$ of an s orbital and $1\frac{1}{2}$ p orbitals are used to form the two C—F bonds (and the two C—H bonds) leaving each carbon atom with $\frac{1}{2}$ of an s-orbital and $\frac{1}{2}$ of a p orbital from which to construct the C—C σ-bond. In the absence of substituents the C—C σ-bond would have been constructed from two sp^2 hybrids each having $\frac{1}{3}s$ and $\frac{2}{3}p$ character.

2

THE FACTORS AFFECTING ATOMIC AND MOLECULAR GEOMETRY: (II) THE STERIC FACTOR

INTRODUCTION

The states of atoms in molecules and the possible electronic interactions of these states are fundamental factors in determining molecular geometry. However, other factors may operate to affect the fine structure of a molecular geometry; one such factor is called the steric factor.

THE STERIC FACTOR

The development of the concept of the steric factor requires an understanding of the concepts of covalent and van der Waals radii.

COVALENT RADII

Atoms having unpaired electrons may bond to form valency saturated molecules containing no unpaired electrons. The bonded atoms are at characteristic separations, called the 'bond length', and at these separations the operative nuclear and electronic attractions and repulsions of the atoms equilibrate. A simple picture of a covalently bonded molecule, A—A, is that of two spherical atoms in contact and having radii equal to a half of the bond length (i.e. a half of the internuclear distance). These radii are called 'covalent radii' (Pauling and Huggins; Schomaker and Stevenson[67]) and they depend on whether the atoms are singly or multiply bonded but for a particular valency state they are relatively constant and independent of the actual molecule. Thus, the experimental value of the bond lengths in molecules A_2 and B_2 gives the covalent radii R_A and R_B; the sum $R_A + R_B$ then gives the bond length in the molecule AB. The covalent radii of the elements may be deduced from a limited amount of experimental data and such data provides a means of calculating bond lengths in all molecules. The covalent radii of the more important atoms found in organic molecules are given in Table 2.1.

Table 2.1. The Covalent Radii of Atoms

Bond	H	C	N	O	P	S	F	Cl	Br	I
Single	0·37	0·77	0·74	0·74	1·10	1·04	0·72	0·99	1·14	1·33
Double		0·67	0·60	0·55	1·00	0·94	0·54	0·89	1·04	1·23
Triple		0·60	0·55	0·50	0·93	0·87				

The constancy of covalent radii is expected, in the absence of other factors, since the bonding of two atoms depends only on the atomic states. However, other factors can operate affecting the values of covalent radii and making molecules very specific geometrical entities. Ionic contributions to a resonance hybrid affect covalent radii and the calculation of R_{AB} requires a modification of the simple additive equation $(R_{AB} = R_A + R_B)$ (Schomaker and Stevenson). Electronic interactions in conjugated systems are also important effects and the importance of these to the covalent radius of carbon is reflected in the concept of 'bond order' (Coulson[68]). The order of a carbon-carbon bond in a molecule is defined as the sum of (a) a contribution by the σ-bonds and (b) a contribution by the π-bonds.

I II III

IV

The contribution is unity for a single bond and the π-bond contribution (the mobile bond order) can be calculated for a particular case; the bond orders for the carbon–carbon bond in ethane, ethylene and acetylene are 1, 2 and 3 respectively. The bond orders for some bonds in benzene, naphthalene, anthracene and phenanthrene are as shown in (I)–(IV) respectively. The bonds in benzene are clearly more nearly ethylenic than ethanic and in the other hydrocarbons some bonds are nearer to double than others. These molecules illustrate the definite effects of conjugation on covalent radii. The relationship between bond order and bond length may be represented graphically and from the curve the bond length may be obtained from the calculated bond order (or vice versa). In spite of the effects of electronic interactions and other factors the concept of covalent radii is a useful one and indeed may be used to detect resonance. Thus, the vinyl chloride molecule, $CH_2=CH-Cl$, has a calculated C—Cl length of 1·76 Å but structures such as $\bar{C}H_2-CH=\overset{+}{C}l$ contribute to the hybrid and the bond is shortened to 1·69 Å.

VAN DER WAALS RADII

The quantum mechanics shows (London[69]) that valency saturated molecules experience a weakly attractive interaction when they are at large separations. The theory further shows (Born[70]) that, as molecules approach, repulsive forces become important and at a certain separation the attractive and repulsive forces equilibrate; at a closer approach than equilibrium strong Born repulsion operates. Molecular interactions of these types were recognized early (e.g. van der Waals[71]); the London forces are in part responsible for the cohesion of gases, liquids and solids. Other types of van der Waals forces can exist but unlike London–Born forces these depend on the presence of certain molecular properties; the nature and origin of

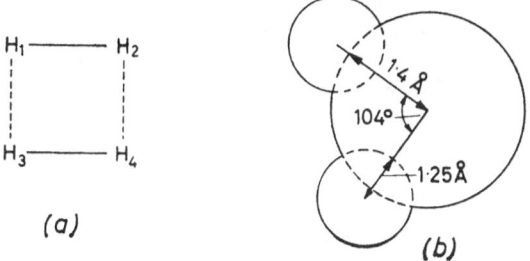

Figure 2.1

London–Born forces need not be discussed at this point. The equilibrium distance (distance of closest approach) of non-bonded atoms in two molecules always exceeds the sum of the covalent radii of the atoms; if molecules could be made to approach closer than this normal minimum they would lose their identity and form a new but unstable species. Thus in *Figure 2.1a* two hydrogen molecules H_1—H_2 and H_3—H_4 have approached to a distance equal to the sum of their covalent radii, this new system is the unstable entity H_4.

Pauling assumed that the distance of closest approach for two atoms in different molecules would be found in the crystalline state. Some variation in this distance is expected when the substance is in the liquid state but this is usually small since little change of density occurs with the change of state; experimentally the internuclear separation of two non-bonded atoms in crystalline bromine is found to be 3·9 Å; a half of this distance is called the van der Waals radius of the bromine atom and in a similar way radii are allotted to other atoms. The sum of the van der Waals radii for any two non-bonded atoms in molecules clearly gives their distance of closest approach i.e. the separation of their nuclei at closest approach. The van der Waals radii for some important atoms are given in Table 2.2; these radii, like covalent radii, derive from a limited

Table 2.2. The van der Waals Radii of Atoms

H	1–1·4 A	S	1·85	I	2·15
N	1·5	F	1·35	CH_3 and CH_2	2·0
O	1·4	Cl	1·8	Aromatic nucleus	1·7
P	1·9	Br	1·95	C	1·5

selection of substances in the crystalline state and are found to be only approximate for general use. Thus, the values are frequently too high, particularly for gaseous and liquid substances, and are generally reduced by 10 per cent. However, the van der Waals radius is quite distinct from the covalent radius and non-bonded closest approach is always a greater distance than bonded closest approach. The van der Waals radius for hydrogen is the most variable which is unfortunate since the outer parts of most organic molecules largely consist of hydrogen. A radius as low as 0·6 Å (intermediate between the covalent radius, taken as 0·3 Å, and the van der Waals radius, taken as 1·2 Å) has been used for the hydrogen atom. The variability of hydrogen has led to the use of an average value for the CH_2 and CH_3 groups. The aromatic nucleus with its π-orbitals perpendicular to the ring plane has a total thickness of 3·4 Å and therefore a 'radius' of 1·7 Å. The data of Pauling (Table 2.2) must also be reduced when considering atoms bonded to the same atom, thus in CCl_4 Pauling's radii suggest that the chlorine atoms overlap with the development of repulsion but frequently these repulsions are rather smaller than expected. In summary, when atoms in different molecules are separated by a distance greater than the sum of their van der Waals radii, then, weak, attractive (London) forces operate, if the separation is less than this sum then powerful (Born) repulsion exists. A valuable and fairly complete stereochemical representation of molecules can be made on the basis of the above concepts. Thus, given the internuclear distances and bond angles in a molecule a skeletal model may be constructed; the model is completed by representing the atoms as spheres of radius equal to the van der Waals radius. The result is the well known Stuart[72] model of partially fused spheres. Planar molecules or groups of atoms lying in a plane may be represented on paper, thus for the water molecule the internuclear distances, H—O, are drawn on an appropriate scale and at the appropriate HÔH. The nuclei are taken as centres and the atoms are then drawn as circles of radius equal to the appropriate van der Waals radius (*Figure 2.1b*) and on the scale used for the internuclear distances. Tables of covalent and van der Waals radii permit therefore, a useful, if approximate, means of representing the geometry of molecules.

The above discussion of van der Waals radii has concerned intermolecular interaction, actually, it is 'intramolecular' interaction which is of primary stereochemical importance. Non-bonded intramolecular interaction follows the principles for intermolecular interaction and attraction or repulsion can operate between two atoms in the same molecule depending on their separation. It is this intramolecular non-bonded interaction which has been called the steric factor and its detailed stereochemical significance can be discussed now for the various groups of organic molecules.

SATURATED ALIPHATIC MOLECULES

The methane molecule has the nuclei of any two hydrogen atoms about 1·78 Å apart; these atoms have low van der Waals radii since they are joined to the same carbon atom and using a radius of 0·75 Å the hydrogens are well clear. However, in tetramethyl methane the large methyl groups experience van der Waals overlap and internal repulsion potential energy exists. This non-bonded repulsion energy is one particular form of 'steric strain energy' and is called 'compression energy'. This strain energy can be relieved, in this particular molecule, only by the stretching of covalent bonds and it is an important stereochemical principle that bond stretching requires much energy. Thus, although a little relief of steric strain may be had by bond stretching, the molecule will have residual compression energy. The existence of compression energy in molecules is of considerable chemical significance (Brown[73]). Steric strain energy originates from Born forces and as such, differs from Baeyer strain although this latter type of strain (deformation of angles) can have a steric cause in certain molecules.

The stereochemistry of ethane has been described previously as that of two linked sp^3 carbon units having slight hyperconjugative interaction; the inclusion of the steric factor is necessary in order to complete the discussion of this molecule. The ethane molecule may be represented in two extreme configurations (*Figures 2.2a* and *b*)

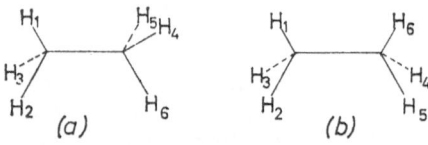

Figure 2.2

related by rotation about the C—C bond. Alternative representations of *Figures 2.2a* and *b* are the projection or 'star' diagrams *Figures 2.3a* and *b* respectively. These latter diagrams are drawings

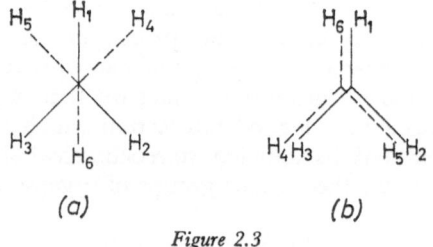

Figure 2.3

of the molecules (*Figures 2.2a* and *b*) as seen by looking along the carbon–carbon bond; the particular configurations *Figures 2.2a* and *b* are called the 'staggered' and 'eclipsed' forms respectively. The staggered form has three H $\cdots$ H distances of 3·06 Å (i.e. $H_1 \cdots H_6$, $H_3 \cdots H_4$ and $H_2 \cdots H_5$) and six H $\cdots$ H distances of 2·49 Å (i.e. $H_1 \cdots H_4$, $H_1 \cdots H_5$, $H_2 \cdots H_4$, $H_2 \cdots H_6$, $H_3 \cdots H_5$ and $H_3 \cdots H_6$); in the eclipsed form three H $\cdots$ H distances of 2·27 Å and six of 2·89 Å exist. If a van der Waals radius for hydrogen of 1·25 Å is assumed it follows that the staggered form has six pairs of hydrogen atoms at about van der Waals contact and the other three interactions must be weakly attractive since the hydrogen atoms are well clear. Thus, the staggered form is stable since no steric compression strain exists. The eclipsed form is clearly less stable than the staggered form since three pairs of hydrogens strongly repel being only 2·27 Å apart; the six distances of 2·89 Å will lead to weak attractions. However, the strained eclipsed form differs in an important way from strained tetramethylmethane in that the compression energy of eclipsed ethane can be relieved completely by the rotation of a methyl group about the carbon-carbon bond to give the staggered structure. Thus the sp^3—sp^3 model of the ethane molecule requires the modification that rotation about the carbon–carbon bond is not entirely free since it involves passing through the eclipsed form and this requires the supplying of energy. These matters may be illustrated in more detail by considering the change of repulsion energy for a complete rotation of a methyl group about the carbon–carbon bond. The rotation of a methyl group of the staggered form (zero repulsion energy) results in a decrease of the 2·49 Å distance below van der Waals contact and repulsion exists. The repulsion potential increases to a maximum value at 60 degree rotation (*Figure 2.4*) when the eclipsed form obtains. A further rotation

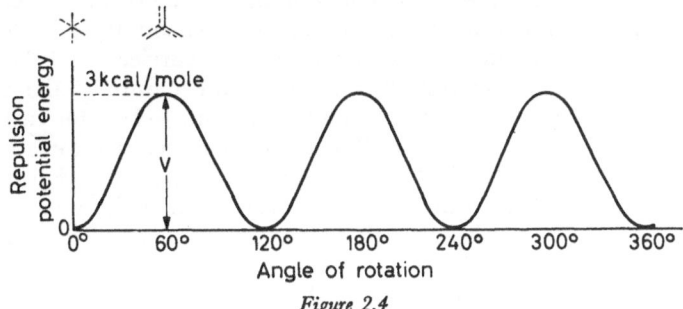

Figure 2.4

reproduces, at 120 degrees, a staggered form of zero energy and the complete 360 degrees involves three staggered forms of minimum energy and three maximum energy eclipsed structures. Thus, in ethane the staggered form is preferred energetically and an energy barrier, of height V exists to hinder free rotation. The barrier in ethane does not prevent rotation at ordinary temperatures but only inhibits it so that at any time most molecules of a mass of ethane will be found in the staggered form; the eclipsed form is the least likely configuration.

An experimental value of the barrier height in ethane is 2·75 kcal/mole (Kistiakowski, Lacher and Stitt[74]); a value as high as 3·3 kcal/mole has been reported; the best present value is 2·875 ± 0·125 kcal/mole (Pitzer[75]). The theoretical aspects of rotation in ethane are not covered entirely by a consideration of the steric factor since invoking only this factor the calculated barrier height is 0·36 kcal/mole (Eyring[76]). Other factors such as electrostatic factors contribute to the energy barrier and more recently (Pauling[77], Eyring[78]) a detailed examination of the importance of hyperconjugation has been made. In spite of the absence of a complete theory the contribution of non-bonded repulsion to the energy barrier is real and stereochemically important.

Any point on the rotation curve (*Figure 2.4*) defines a configuration of atoms and up to the first 60 degrees of rotation these are different; these different geometries are called 'conformations' (Haworth[79]). A conformation may be defined more precisely as any molecular geometry which can arise from internal rotations or vibrations at ordinary temperatures. The staggered and eclipsed ethanes are clearly conformations. It will be clear from *Figure 2.4* that an infinite number of conformations is possible for a molecule but only the stable ones (those of minimum energy) are of stereochemical interest and these are limited in number. Conformations related by internal rotations are the more usual ones encountered in stereochemistry. The existence of an internal steric barrier is not the sole criterion determining restricted rotation, the nature of the bond about which rotation occurs is also important. Thus, for carbon–carbon bonds, rotation is possible at ordinary temperatures if the bond order is unity. Rotation is frequently possible, depending on the molecule, at ordinary temperatures, when the bond order exceeds unity but it is impossible for a bond order 2. The limiting case of bond order 3 also permits rotation at ordinary temperatures. One concludes that a substance made up of molecules which, due to a steric energy barrier and/or a particular bond order, can exist as different conformations is to be regarded as a labile mixture of flexible geometries with the most stable form in preponderance.

The steric factor also operates in *n*-butane and if the molecule is regarded as an ethane molecule substituted by two spherical methyl groups (radius 2 Å) the rotation of interest is that at about the central carbon–carbon bond. The four extreme conformations (*Figures 2.5a, b, c* and *d*) occur during the rotation and the potential energy curve for a complete rotation is shown in *Figure 2.6*. An important difference between ethane and *n*-butane is that the

C—CH_3 bonds are longer (1·54 Å) than the C—H bonds (1·1 Å) of ethane which they replace, hence, the replacement of two hydrogen atoms of staggered ethane by the larger methyl groups does not lead to any new van der Waals repulsion. The two methyl groups (*Figure 2.5a*) are at about van der Waals contact as are the methyl

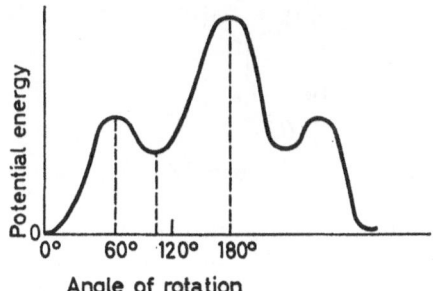

Figure 2.5

groups with the hydrogen pairs H_1, H_2 and H_3, H_4. A rotation of half of the staggered conformation about the central carbon-carbon bond results in the development of repulsion between a CH_3 and H_1, a CH_3 and H_4 and between $H_2 \cdots H_3$. A repulsion must develop also between the two CH_3 groups but this will be small particularly in view of the compressibility of alkyl groups; at 60

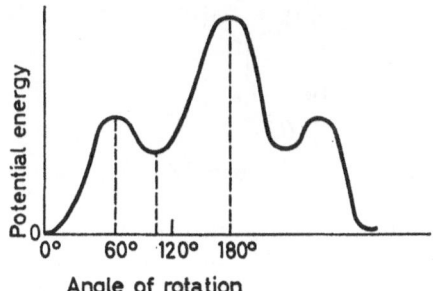

Figure 2.6

degrees rotation the eclipsed form (*Figure 2.5b*) exists and the repulsion energy, due to $CH_3 \cdots H_1$, $CH_3 \cdots H_4$ and $H_2 \cdots H_3$ is a maximum; the $CH_3 \cdots CH_3$ interaction in this conformation is small and, as will be shown later, may be discounted. On further rotation the $CH_3 \cdots H$ and $H \cdots H$ interactions decrease and disappear at 120 degrees; however, a rotation beyond 60 degrees increases the weak $CH_3 \cdots CH_3$ interactions and the minimum of energy is at slightly less than 120 degrees; this is the skew form (*Figure 2.5c*). The potential energy maximum at 180 degrees rotation corresponds to the fully eclipsed conformation (*Figure 2.5d*) and further rotation reproduces the skew, eclipsed and staggered forms. The steric strain energy differences between the staggered form and the other conformations are, for the eclipsed form 2·9 kcal/mole, for the skew form 0·8 kcal/mole and for the fully eclipsed form 3·6 kcal/mole (Pitzer[80]).

Thus, n-butane can exist in two stable conformations (staggered and skew); their interconversion via the eclipsed form requires 2·9 kcal/mole and, via the fully eclipsed form, 3·6 kcal/mole is required. The above conformational analysis is superficial and in error in certain detail since the CH_3 groups were regarded as spherical. In a detailed analysis, using the complete tetrahedral structure, a variety of staggered forms can be drawn, for example, *Figures 2.7a* and *b*. The form, *Figure 2.7a*, has the C_2—C_3 ethane

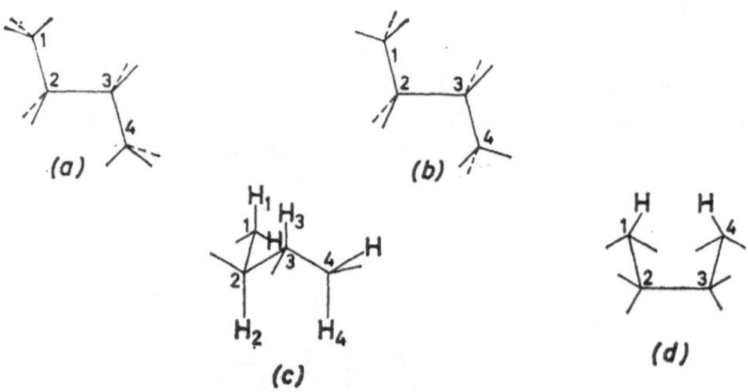

Figure 2.7

unit staggered but the C_1—C_2 and C_3—C_4 units are eclipsed; the form *Figure 2.7b* has staggered units throughout and is clearly the most stable of the possible staggered conformations. The conformation *Figure 2.7b* represents that of zero energy in *Figure 2.6*. In fully staggered forms no other interactions are possible, thus, in *Figure 2.7b* the replacement of an ethane hydrogen by the larger carbon atom C_1 might suggest interaction between this carbon and the hydrogens on C_3. However, the increased size of carbon compared with hydrogen is more than offset by the increased bond lengths, C_1—C_2 = 1·54 Å while C—H = 1·1 Å. The conformation *Figure 2.7c* is a detailed skew form and is the most stable one since the ethane units are staggered. However, the geometry of skew n-butane introduces the new interaction between the unnumbered hydrogens on C_1 and C_4. These hydrogens are at about 1·9 Å apart and repulsion leads to a slight twist of the C_2—C_3 bond to give the <120 degree minimum energy conformation; the residual repulsion amounts to 0·8 kcal/mole. The separation of the numbered hydrogens on C_1 and C_3 and C_2 and C_4 of skew n-butane is also an important distance; the separation is 2·51 Å and results in the H_1, H_3 and H_2, H_4 hydrogens shown, being in van der Waals contact. The replacement of 1,3-hydrogens by larger groups can lead to repulsion and this form of interaction is important in certain cyclohexane derivatives. The 1,4-skew n-butane interaction is important in certain polycyclic systems. The eclipsed form *Figure 2.7d* has staggered ethane units C_1—C_2 and C_3—C_4 but the hydrogens shown on C_1 and C_4 are only about 1 Å apart and strongly interact; 1,4-interactions of this type are also important in cyclohexane stereochemistry.

Higher aliphatic hydrocarbons are similar to n-butane; the fully staggered form is the most stable one; n-pentane has two preferred conformations while n-hexane and n-heptane have three preferred forms. The numbers of stable forms for higher hydrocarbons have not been defined as yet. The steric energy barriers separating the stable forms of simple substituted ethanes, such as ethylene dichloride, are about the same order as in n-butane and since the bond order of the central C—C bond in these substances is about 1 they exist, like n-butane, as labile mixtures of conformations at ordinary temperatures. Low barriers are found also in other simple aliphatic systems; in CH_3OH the non-bonded interaction of the hydroxylic and methyl hydrogens gives a barrier of 0·932 kcal/mole. Steric barriers are found in CH_3NH_2 (3 kcal/mole), acetone (1 kcal/mole) and acetaldehyde (1 kcal/mole). The CH_3NO_2 molecule is of special interest since when one nitro group oxygen atom is fully eclipsed by a methyl hydrogen atom the other oxygen atom is staggered between two hydrogen atoms; thus, the steric factor is the same for all conformations.

The steric factor can operate also in combinations of states such as sp^3—$\overset{+}{C}$, sp^3—$\overline{C}$ and sp^3—C·, but with rather different effects. The ion CH_3—$\overset{+}{C}H_2$ will be most stable as conformation (*Figure 2.8a*); some repulsion exists between the hydrogen atoms on the different carbons but this is insufficient to distort the planar carbonium ion. The methyl radical is equally stable in pyramidal or planar form but the planar ethyl radical would have a geometry like that of *Figure 2.8a* and so the hydrogen repulsions favour the pyramidal form *Figure 2.8b*. The pyramidal radical (*Figure 2.8b*) is

E .

3·5 kcal/mole more stable than the planar form (Skell, Woodworth and McNamara). The ethyl carbanion is very similar to ethane but the lone pair probably has a greater steric effect than a hydrogen atom; the ion has the staggered form.

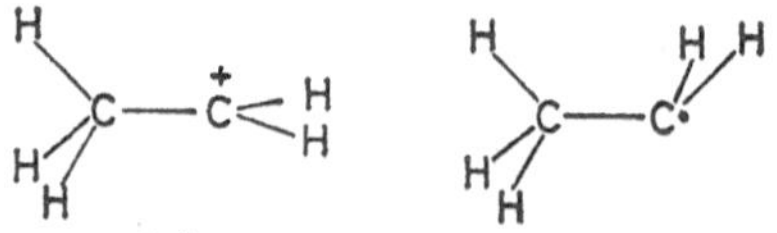

(a)   *(b)*

Figure 2.8

ALICYCLIC MOLECULES
Monocyclic Hydrocarbon Molecules

In a strained conformation of an aliphatic hydrocarbon, for example, eclipsed ethane or skew n-butane, the steric factor may be relieved completely by a rotation about a σ-bond. The position in certain alicyclic molecules is different, thus the cyclopropane molecule has eclipsed hydrogen atoms and internal repulsion exists. However, unlike eclipsed ethane the repulsions are tolerated because their relief would entail rotation about C—C bonds and this would reduce the overlap of the already bent bonds so leading to further destabilization. The cyclopropane molecule has, therefore, three sources of strain, namely the strain of closing the tetrahedral angle to a value of 106 degrees ('angle strain'), the overlap strain resulting from non-linear overlap of bonding orbitals, and the repulsion strain of the eclipsed hydrogens. The barrier to rotation in ethane is about 3 kcal/mole, hence each pair of eclipsed hydrogens in a molecule contributes a strain of about 1 kcal/mole; in cyclopropane this strain amounts to about 6 kcal/mole. The cyclobutane molecule actually is twisted, the twisting decreases the steric strain from 8 kcal/mole to 5 kcal/mole. Actually the geometry of cyclobutane introduces a new source of steric strain due to the interactions of carbon atoms at opposite corners; their separation is about 2·2 Å and is well below their van der Waals sum of 3 Å.

Classical stereochemistry regarded alicyclic molecules as having planar rings and introduced the concept of angle strain, in classical form, to account for the planarity. A planar regular pentagonal cyclopentane molecule has $\widehat{CCC}$ = 108 degrees. This angle is so close to the tetrahedral angle that the molecule may be assumed to be constructed from essentially sp^3 carbon atoms; it will have a strain of 10 kcal/mole due to eclipsed hydrogens. In this molecule the relief of non-bonded interaction is possible by some form of internal rotation, thus, if in the planar model *Figure 2.9a* the methylene groups C_2 and C_5 are rotated about the C_2—C_3 and C_5—C_4 bonds respectively then the C_2—C_1 and C_5—C_1 bonds of the methylene groups will leave the ring plane and, in order to maintain

(a)　　　　　　(b)　　　　　　(c)　　　　　　(d)

Figure 2.9

orbital overlap, the C_1 group must also leave the original plane. The puckered molecule (*Figure 2.9b*) results (Aston[81]). The molecular puckering has the important effect of decreasing the internuclear angles $C_4\widehat{\,}C_5\,C_1$, $C_5\,\widehat{C_1}\,C_2$ and $C_1\,\widehat{C_2}\,C_3$; in turn this results in strain due to bent bonds as illustrated for the $C_5\,\widehat{C_1}\,C_2$ in *Figures 2.9c* and *d*, the broken lines in *Figure 2.9d* show the bent bonds. The distortions in cyclopentane are slight and are probably met by a change of hybridization rather than the bending of bonds. Thus, planar cyclopentane cannot be constructed from sp^3 carbons (without bent bonds) but a planar molecule may be derived by rehybridizing the ring carbon sp^3 orbitals to sp^x ($x > 3$) so that linear overlap occurs and the ring angles are 108 degrees. This use of sp^x orbitals produces an angle strain of 0·1 kcal/mole. The puckering requires still further rehybridization if σ-bonds (and not bent bonds) are to be retained; the orbitals will be sp^y ($y > x$) so introducing more angle strain. The relief of non-bonded strain by puckering stabilizes the molecule but simultaneously it introduces destabilization due to angle strain; at some point the stabilization and destabilization equilibrate and the rotation and puckering will cease. The relative changes of various types of internal strain with a change of geometry is a matter of great stereochemical importance. Thus non-bonded repulsion falls off very rapidly with increased separation of the overlapping atoms and a slight puckering considerably decreases this form of strain. The angle strain developed by puckering is far less than the decrease of non-bonded repulsion making the puckered form the more stable conformation. At equilibrium increased puckering or increased planarity will both lead to an increase of strain energy relative to that of the normal puckered form. The strain energy of the planar form is 10·1 kcal/mole, clearly the sum of residual repulsion and angle strain in the puckered form will be less; experimentally it is found to be 6 kcal/mole.

The energy difference between planar and puckered cyclopentane is small and, at room temperature, they are easily interconverted. The planar form may pucker so that the displaced carbon atom becomes either above or below the plane of the other four carbons. Thus, the above and below plane puckered conformations are readily interconverted. This interconversion is called 'inverse puckering' and although for cyclopentane it is unimportant, because the above and below plane forms are identical, it is important for substituted cyclopentanes. The above discussion has considered the displacement of a particular carbon atom but the energetics of puckering is such that, in a particular molecule, each carbon atom in turn can become displaced above or below the ring plane by internal rotations of appropriate methylene groups in the planar form. The process is called 'mobile puckering'.

The classical Baeyer model of cyclohexane is that of an angle strained planar, regular hexagon having $\widehat{CCC} = 120$ degrees. This model can be constructed, in principle, from six carbon atoms each having two sp^2 orbitals with which to form the ring; the carbon orbitals carrying the two hydrogen atoms will be equivalent and have about 16 per cent s- and 84 per cent p-character. Alternatively the planar geometry may be constructed from sp^3 carbons but it will have bent bonds. Thus, the sp^2 model has angle strain (of C—H and C—C bonds) and the sp^3 model has overlap strain (of C—C bonds); both have the steric strain of eclipsed hydrogens. Actually these highly strained planar models unlike the case for

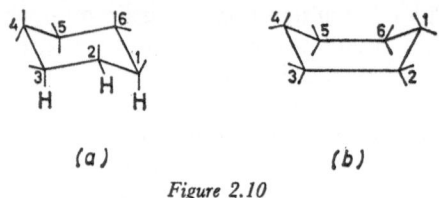

(a)   (b)

Figure 2.10

cyclopentane, have no reality even as unstable intermediates. The geometry of cyclohexane is, in fact, a simpler matter than for cyclopentane since it is found that six sp^3 carbon atoms can be joined, with linear overlap of the ring carbon orbitals, as a puckered collection of staggered ethane–skew n-butane units (*Figure 2.10a*); this geometry is called the 'chair' structure. This structure has the two groups of carbon atoms, $C_1 C_3 C_5$ and $C_2 C_6 C_4$, situated in parallel planes at 0·5 Å separation. Further, these groups of carbon atoms are at the apices of two identical equilateral triangles, a fact difficult to make clear in a drawing.

As pointed out the skew hexagonal 'chair' form consists of six staggered ethane units and hence the 2·49 Å H · · · · H distance is found. The separation of the hydrogens on C_1 and C_2, for example, is such a distance. However, the number of these distances is reduced to three per ethane unit (ethane itself has six 2·49 Å distances) because in cyclohexane some of the hydrogens of the ethane units from which it is constructed are replaced by carbon atoms. The total number of 2·49 Å H · · · · H distances in cyclohexane is eighteen, the cyclohexane molecule also contains skew n-butane units and as shown in *Figure 2.10a* the 1,3-H · · · · H separation of 2·51 Å is found but the 1,4-distance does not occur since the hydrogen atoms are replaced by carbons and these do not interact. The 2·51 Å H · · · · H distance occurs six times in cyclohexane. The construction of a cycle of ethane units, in 'chair' form, does not introduce any further non-bonded interactions or potential interactions. Thus, in the 'chair' form twelve pairs of hydrogen atoms are at about van der Waals contact of 2·49 Å or 2·51 Å; these may be classified as 1,2- or 1,3-pairs.

Any internal rotations about the carbon–carbon bonds of the cyclohexane molecule will result in a decrease of the 2·49 Å distance (similar rotations increase the 2·51 Å distance) and 1,2-repulsions are introduced; the existence of eighteen 1,2-contacts clearly confers rigidity and stability on the 'chair' geometry.

A second non-planar form of cyclohexane, containing σ-bonds and tetrahedral angles, can be constructed from sp^3 carbon atoms (*Figure 2.10b*); this is called the 'boat' form. The 'chair' form has no intramolecular strain but the 'boat' form is strained since it contains eclipsed ethane units. Thus, the hydrogens on the 'sides' of the 'boat' (C_2—C_3 and C_5—C_6) are eclipsed. The 'boat' form also contains eclipsed n-butane units and the 'prow' and 'stern' hydrogens, being at 1·83 Å, experience the strong 1,4-interactions described previously. The hydrogen atoms of the ethane units C_3—C_4, C_4—C_5, C_6—C_1 and C_1—C_2 are staggered and make no contribution to the strain energy. The repulsion of the 1,4-hydrogen atoms cannot be determined by analogy with eclipsed n-butane since aliphatic systems can adjust themselves to some extent sometimes when internal strains are set up; adjustment is less easy in cyclic systems. However, the strain energy of the 'boat' form must exceed that of the 'chair' form by at least 4 kcal/mole due to eclipsed hydrogen interactions. Several estimates of the total strain difference between the 'chair' and 'boat' forms have been made, the value is 5·3 kcal/mole (Johnson[82]). The reality of the stable 'chair' geometry has been established by physical evidence (Hassel[83]).

The 'chair' and 'boat' forms have been considered so far merely as possible structures for cyclohexane, actually they are conformations since they may be interconverted by internal rotations and the energy difference of 5·3 kcal/mole is attainable at ordinary temperatures. The rotational interconversion of 'chair' and 'boat' forms is best seen by drawing *Figure 2.10a* as the projection *Figure 2.11a*; this diagram corresponds to a view of the molecule as seen by

(a)   *(b)*

Figure 2.11

looking along the C_2—C_3 and C_6—C_5 bonds; the hydrogen bonds on C_1 and C_4 are omitted in *Figure 2.11a*. A rotation of the methylene groups C_2 and C_6 about the C_2—C_3 and C_6—C_5 bonds, anti-clockwise and clockwise respectively, results in the partial eclipse of hydrogens on C_1, C_2 and C_6; after a 30 degree rotation these atoms lie in the plane of the C_2—C_3 and C_5—C_6 bonds. Further rotation results in a further increase in the repulsion energy and this becomes maximum when the 'boat' form (*Figure 2.11b*) is produced after a 60 degree rotation. Although the 'chair' and 'boat' forms are interconvertible, physical evidence shows that only trace amounts of the 'boat' form can be present in cyclohexane.

In addition to the 'chair'—'boat' interconversions a second type of rotational interconversion exists. Thus, the 'chair' form (*Figure 2.10a*) may be converted to the 'boat' form (*Figure 2.10b*) as described; internal rotation may now convert the 'boat' form back to the original 'chair' form or, by a rotation of groups C_3 and C_5, the inverse of the original 'chair' is obtained. This inverse puckering, via, the 'boat' form, is unimportant for cyclohexane itself, since the two 'chair' forms are identical, but it becomes of great importance in the stereochemistry of substituted cyclohexanes. The 'boat' form is unstable and so internal rotations easily occur; a succession of rotations can make each carbon atom the 'prow' of the 'boat' in turn. The 'boat' conformation is therefore flexible and undergoes mobile puckering.

The above discussion of the conformations of cyclohexane has considered only the steric factor. However, as in ethane, other factors may affect stability but the physical evidence shows that they operate in the same direction as the steric factor, at least in that the steric factor alone can predict satisfactorily the preferred conformation. Actually, the 'chair' and 'boat' models were recognized quite early (Sachse[84]; Mohr[85]). It was appreciated that these models, constructed from tetrahedral carbon atoms, were superior to Baeyer's in that no angle strain exists. However, proof of their reality and an interpretation in terms of the steric factor is of recent origin.

The problem of preferred conformations for higher alicyclic hydrocarbons is of current interest. It appears that the preferred forms of medium size rings (8–12 carbons) contain some eclipsed hydrogens but all other rings have the usual staggered form (Prelog[86]).

Simple Substituted Monocyclic Molecules

Substituents in an alicyclic system will affect the states of hybridization of the atoms slightly but these effects can be ignored relative to the steric effects which result.

An examination of the carbon-hydrogen bonds in 'chair' cyclohexane (*Figure 2.12a*) shows that they fall into two geometrically different groups (Kohlrausch[87]; Hassel; Pitzer[88]). These groups are separately shown as *Figures 2.12b* and *c*. One group of six bonds (*Figure 2.12b*) is parallel to a molecular axis and they are called axial bonds; the second group (*Figure 2.12c*) radiates from the ring and lies in the general plane of the ring, they are called equatorial bonds. The existence of these two groups of bonds immediately raises a stereochemical problem for substituted cyclohexanes since the bromine atom in bromocyclohexane, for example, may be attached either axially or equatorially and so exist in two different geometries.

Any axial hydrogen, for example that on C_1 (*Figure 2.12a*) is in van der Waals contact with the equatorial hydrogens on C_2 and C_6 (1a–2e = 1a–6e = 2·49 Å) and also with the axial hydrogens on C_3 and C_5 (1a–3a = 1a–5a = 2·51 Å). The replacement of this axial hydrogen by the larger bromine atom results in considerable repulsions between bromine and hydrogens since the sum of the van der Waals radii for bromine (1·95 Å) and for hydrogen (1·25 Å) is 3·2 Å which is rather larger than the 2·49 Å and 2·51 Å distances. Actually, the centre of the bromine nucleus does not coincide exactly with that of the replaced hydrogen nucleus since the C—H and C—Br bond lengths are 1·1 Å and 1·93 Å respectively. This results in an increase of the H · · · · Br internuclear distances

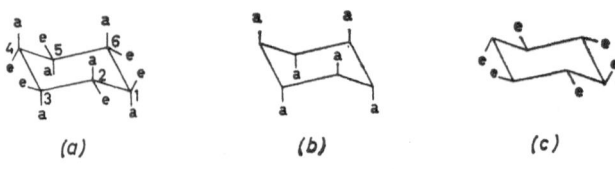

(a)   *(b)*   *(c)*

Figure 2.12

(1a–2e, 1a–6e, 1a–3a, 1a–5a) in the bromo compound to about 2·6 Å (as opposed to 2·49 Å and 2·51 Å in cyclohexane) but this is still well below 3·2 Å. A further effect of the change of bond length· though not obvious, is that if the H · · · Br distances (1a–3a, 1a–5a) are taken as exactly 2·6 Å then the H · · · · Br distances (1a–2e, 1a–6e) are slightly greater than 2·6 Å and so the 1,3-repulsions are greater than the 1,2-repulsions.

An equatorial hydrogen has the van der Waals contacts 1e–2e, 1e–2a, 1e–6e and 1e–6a; the introduction of a bromine atom will result in four 1,2-repulsions. In this case the change of bond length will affect these interactions equally. Thus, an axial bromine atom has two 1,2-interactions and two stronger 1,3-interactions; these latter dominate the energy relations and the bromine atom prefers the equatorial position with the four 1,2-interactions. This preferred conformation is found for monosubstituted cyclohexanes in general although the small fluorine atom appears to be most stable axially;

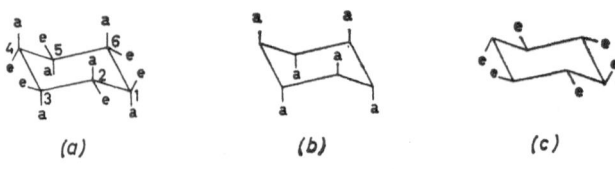

(a)   *(b)*

Figure 2.13

thus the equatorial forms of cyclohexanol and bromocyclohexane are 0·3 kcal/mole and 0·6 kcal/mole more stable, respectively, than the axial form; that equatorially and axially substituted cyclo-hexanes are really conformations may be illustrated by methyl-cyclohexane. In this molecule the axial form is less stable than the equatorial form by 1·8 kcal/mole, further inverse puckering provides a mechanism for the interconversion of the forms. Thus, inverse puckering converts axial bonds into equatorial bonds (and vice versa) as shown in *Figures 2.13a* and *b*.

One concludes that the steric factor results in the 'chair' conformation and equatorial substitution being preferred for cyclohexane derivatives.

Polycyclic Molecules

The properties for maximum conformational stability namely the presence of (*a*) staggered ethane units throughout and of (*b*) equatorial substitution apply to polycyclic systems. Thus *Figures 2.14a* and *b* are possible conformations for dicyclohexyl; *Figure 2.14a* is

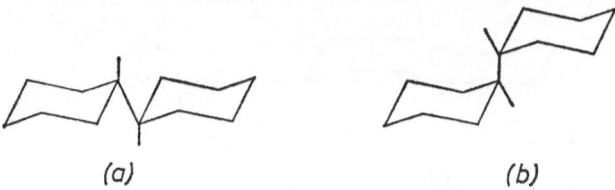

(*a*)   (*b*)

Figure 2.14

clearly the most stable one. The polycyclic substance decahydronaphthalene ('decalin') is of historical stereochemical interest. By analogy with cyclohexane Mohr suggested two possible geometries, *Figures 2.15a* and *b*, for the decalin molecule; a third possibility (*Figure 2.15c*) was pointed out later by Wightman[89] and by Hassel.

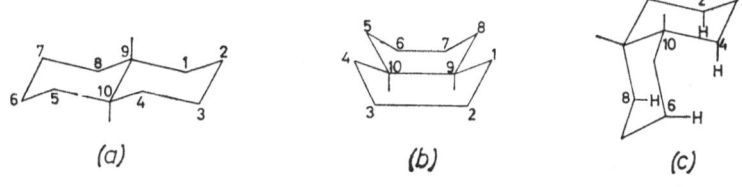

(*a*)   (*b*)   (*c*)

Figure 2.15

The 'chair–chair' form (*Figure 2.15a*) has the two rings fused so that no interactions occur and the molecule is strain free. However, the 'chair–chair' form (*Figure 2.15c*) has a ring fusion such that the hydrogens shown on C_4, C_6, C_2, C_8 and C_8, C_4, experience 1,4-skew n-butane interactions and these produce a steric strain of $3 \times 0.8 = 2.4$ kcal/mole. The 'boat–boat' form, *Figure 2.15b*, has the hydrogens on the 'sides' of the 'boats' eclipsed (C_2—C_3, C_4—C_{10}

and C_6—C_7) and the 'axial' prow hydrogens on C_1 and C_8 interact with 'axial' stern hydrogens on C_4 and C_5 respectively. This structure is found to be 8·8 kcal/mole less stable than *Figure 2.15c* (Turner[90]). That structures *Figures 2.15a* and *c* represent the possible stable geometries for the decalin molecule has been established experimentally (Bastiansen and Hassel[91]); the structure *Figure 2.15b*, is unknown.

The relationships between these structures are further clarified by considering the possible internal rotations. The 'chair–chair' structure (*Figure 2.15a*) is flexible only in part, motion of the parts C_5, C_6, C_7 and C_1, C_2, C_3 is possible with the production of conformation *Figure 2.16a*. The part of the molecule C_8, C_9, C_{10}, C_4 is

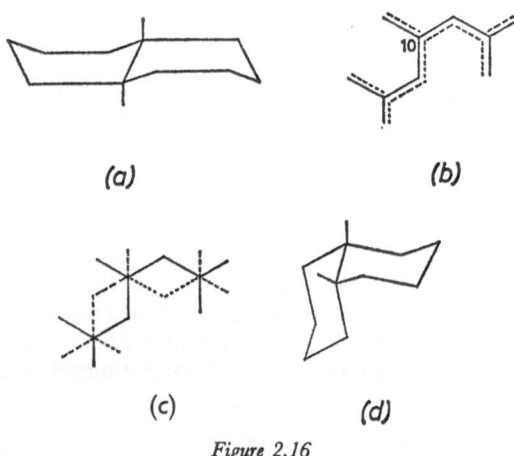

(a)

(b)

(c)

(d)

Figure 2.16

rigid since rotation would convert the equatorial bonds linking the rings into axial bonds and the fusion of cyclohexane rings through adjacent axial bonds is impossible since the ring is too small. The 'boat–boat' form *Figure 2.16a* is, of course, different to that of *Figure 2.15b*. The decalin, *Figure 2.15c*, has an axial and equatorial fusion of the rings and so is flexible, thus, inverse puckering is possible and axial and equatorial bonds are interconvertible throughout the molecule. The interconversion proceeds via the 'boat–boat' form, *Figure 2.16b*; this conformation then proceeds to the 'chair–chair' form *Figure 2.16c*. This latter conformation is easier to relate to *Figure 2.15c* when drawn as *Figure 2.16d*. The intermediate 'boat–boat' form (*Figure 2.16b*) is again different from

Figure 2.15b. One concludes that the decalins *Figure 2.15a* and *c* are not conformations but individual geometries; *Figure 2.15a* is rigid while *Figure 2.15c* is flexible and can exist as two stable conformations neither of which is identical to *Figure 2.15b*.

Other polycyclic systems are also most stable as completely staggered structures; the structures *Figures 2.17a* and *b* represent

(a)   (b)

Figure 2.17

the most stable forms of perhydroanthracene and of the steroid nucleus respectively.

Alicyclic Molecules Containing Other Carbon States

The alicyclic systems so far discussed contain only sp^3 carbon, other states, such as sp^2, sp, $\overset{+}{C}$, $\overset{-}{C}$ or $C\cdot$, can be introduced into the cyclohexane ring.

The sp^2 state—Cyclohexanone is an example; the sp^2 state has little effect on the stereochemistry of the molecule as seen in *Figure 2.18a*. It is of interest that due to the absence of 1,4-interactions the cyclohexan-1,4-dione molecule can exist in the 'boat' form *Figure 2.18b*. Cyclohexene has two adjacent sp^2 atoms and it has the 'half-chair' form, *Figure 2.18c*. The bonds designated a and c are identical to ones in cyclohexane but those designated a' and c' are not truly axial and equatorial but only approximately so.

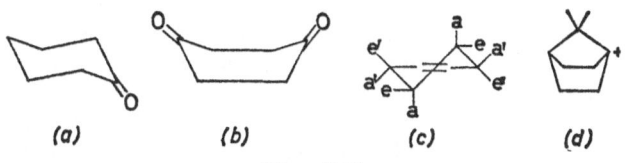

(a)   (b)   (c)   (d)

Figure 2.18

The sp state—The linear sp state is not found in lower alicyclic systems; the steric lower limit for the existence of an acetylenic linkage in a ring is with an eight carbon system (Blomquist[92]). Higher membered rings have also been prepared (Ruzicka[93]).

The $\overset{+}{C}$, $\overset{-}{C}$ and C radical states—A cyclohexyl carbonium ion is conformationally like cyclohexanone, a carbanion is like cyclohexane and the lone pair presumably occupies the equatorial position. The cyclohexyl radical will also have the cyclohexane stereochemistry with the unpaired electron equatorial. Bridge-head structures such as the carbonium ion (*Figure 2.18d*), the carbanion and the radical derived from apocamphane compounds have a 'boat' form. The geometry is rigid and this results in the ions and radical having a tetrahedral geometry.

Alicyclic Molecules Containing Heteroatoms

Cyclohexane has a C—C bond length of 1·54 Å and a $\widehat{CCC} =$ 109 degrees 28 minutes; aliphatic nitrogen and oxygen have the C—N and C—O bond lengths as 1·47 Å and 1·44 Å; the $\widehat{CNC}$ and $\widehat{COC}$ are 109 degrees and 111 degrees respectively. Thus, an NH group or an oxygen atom can replace a CH_2 group in cyclohexane without serious distortion of the skew hexagonal geometry. The tetrahydropyran molecule has the stable conformation *Figure 2.19a*

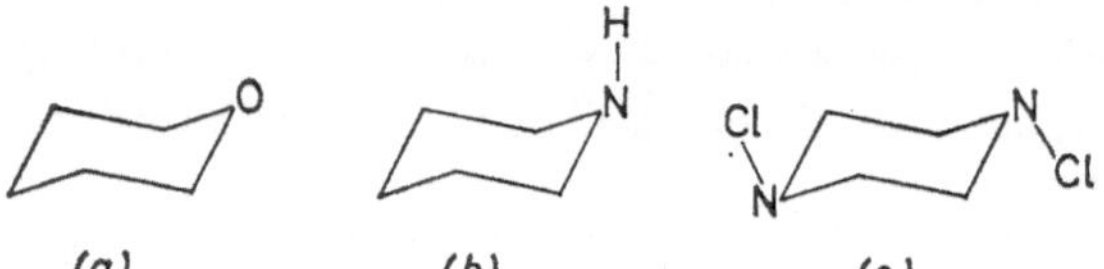

(a)   (b)   (c)

Figure 2.19

and this geometry is found in the pyranose sugars. Piperidine and N,N-dichloropiperazine have the conformations *Figure 2.19b* and *c* respectively; the lone pair on the nitrogen atom is clearly larger than a hydrogen atom but smaller than chlorine since in *Figure 2.19b* it is equatorial while in *Figure 2.19c* it is axial.

UNSATURATED AND CONJUGATED MOLECULES

Rotation about pure double bonds (bond order 2) is possible but these bonds are rather rigid systems and the energy quantities involved far exceed those for rotation about σ-bonds (bond order 1). Thus, rotation about double bonds cannot take place at ordinary temperatures unless other factors operate to reduce the bond order below 2. Conjugation can reduce bond order and these systems contain double bonds with partial single bond character and single bonds with partial double bond character.

Simple Unsaturated Molecules

Ethylene is a planar molecule with its eclipsed hydrogens at about 2·5 Å; clearly no steric factor operates in the molecule. Planar ethylene is the preferred geometry for two linked sp^2 carbon atoms because this configuration minimizes the overlap strain of the p orbitals and maximizes π-bonding. The energy changes for a complete rotation about the double bond of ethylene are shown in *Figure 2.20*. A rotation from the planar configuration results in decreased p orbital overlap and the potential energy increases.

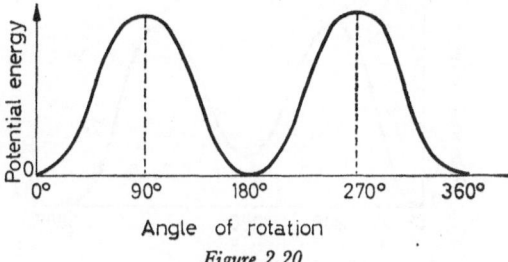

Angle of rotation

Figure 2.20

Although the overlap decreases, it remains considerable even at quite large angles of twist and is destroyed only at 90 degrees rotation; this 'perpendicular' ethylene clearly has maximum overlap strain. Further rotation beyond 90 degrees increases the overlap and decreases the energy to a minimum value at 180 degrees. These energy changes are repeated for the rotation 180 degrees $\rightarrow$ 360 degrees. The overlap strain results in a barrier to rotation of 60 kcal/mole (Mulliken[94]).

The two structures (*Figures 2.21a* and *b*) similarly represent the stable forms of 1,2-dimethylethylene. However, in the structure *Figure 2.21b* non-bonded repulsion occurs between the methyl groups and this form has an energy of 1·28 kcal/mole higher than that of *Figure 2.21a*. The methyl groups in *Figure 2.21a* are at about

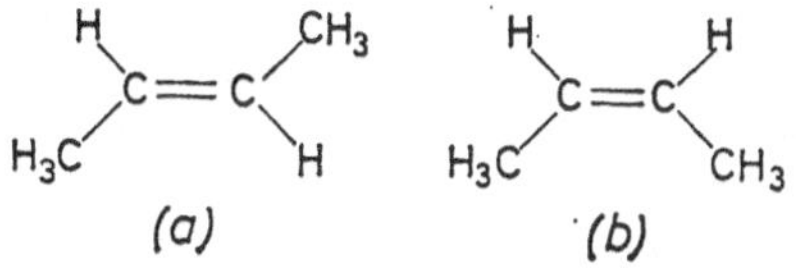

$$\begin{array}{cc}
\text{(a)} & \text{(b)}
\end{array}$$

Figure 2.21

van der Waals contact hence a rotation from the planar configuration results in the development of (a) overlap strain and of (b) non-bonded strain due to methyl group interaction. The overlap barrier is 18 kcal/mole and beyond a 90 degree rotation the decrease in overlap strain easily offsets the continued increase of the steric strain. The overlap strain decreases to a minimum value at 180 degrees while the steric strain increases to the maximum value of 1·28 kcal/mole at this rotation (*Figure 2.22*). Thus, the geometries, *Figures 2.21a* and *b*, represent independent structures and not conformations because the overlap barrier is too high to allow

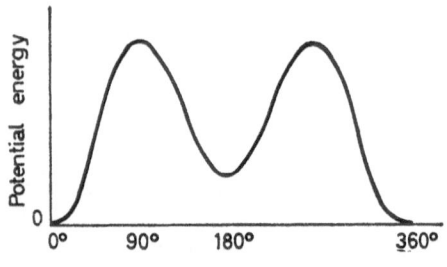

Figure 2.22

interconversion at room temperature. Actually, the height of the barrier depends on the actual conversion being considered, thus, the barrier for *Figure 2.21a → Figure 2.21b* is 18 kcal/mole but for the reverse change it is (18 − 1·28) kcal/mole; in principle it follows that a large enough steric factor can result in only one stable structure existing.

The steric strain in *Figure 2.21b* is small but within an appropriate molecule it can become so large that some relief is necessary. Relief of steric strain in this type of system may be obtained by a slight twist about the double bond and/or the bending of the carbon-methyl group bonds.

Conjugated Molecules

Aliphatic molecules—Quantum mechanics reveals three types of conjugation namely the π-σ-type, found in propylene, the π-p-type, found in vinyl chloride, and the classical π-π-type found in but-1,3-diene. These effects result in classical double bonds in a molecule acquiring partial single bond character and vice versa. Thus, internal rotation about classical double bonds is facilitated and rotation about single bonds is hindered. Any internal rotation leads to non-planarity and interferes with conjugation.

The π-σ-type conjugation, hyperconjugation, is found in the previously discussed 1,2-dimethylethylene. The hyperconjugation present results in the classical double bond having an order less than 2 (the $C{=}C$ bond length is 1·4 Å); this is responsible for the lower barrier height relative to ethylene. The conjugation in vinyl chloride involves the π-bond and a p orbital of the chlorine atom; the double bond overlap barrier is again reduced. The but-1,3-diene molecule may adopt two most stable structures (*Figures 2.23a* and *b*) for rotation about the classical single bond. These configurations are

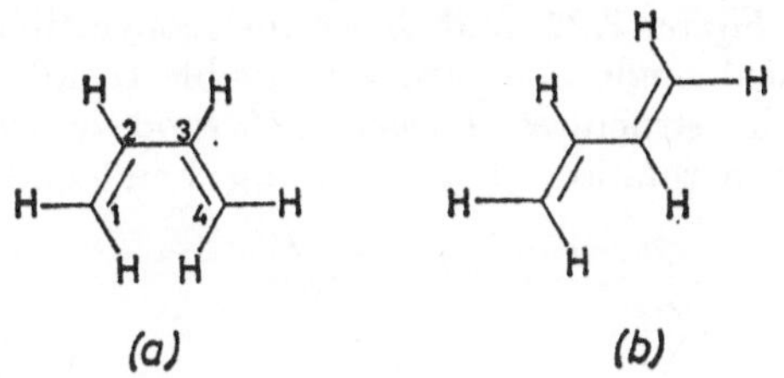

(a)   (b)

Figure 2.23

planar and conjugation is optimum. However, the structure (*Figure 2.23a*) is 2·3 kcal/mole less stable than *Figure 2.23b* due to the steric repulsion between the hydrogen atoms on C_1 and C_4. The double bond character of the classical single bond C_2—C_3 (bond order 1·447, bond length 1·46 Å) creates an overlap barrier of 2·6 kcal/mole and, therefore, a complete interconversion *Figure 2.23b* → *Figure 2.23a* requires (2·6 + 2·3) kcal/mole. One concludes that the geometries, *Figures 2.23a* and *b*, represent conformations and that *Figure 2.23b* is the preferred one.

In the conformation the repulsions could be relieved in part by rotations about bonds. In principle rotation is possible about the C_1—C_2 and C_3—C_4 bonds or about the C_2—C_3 bond. However, the bond orders are C_1—C_2 = C_3—C_4 = 1·894 and C_2—C_3 = 1·447, clearly rotation about the C_2—C_3 bond is the one to occur. A rotation about the C_2—C_3 bond reduces the resonance (destabilization) and reduces the steric factor (stabilization); the rotation ceases when these two factors equilibrate and the resultant conformation is non-planar. The strain energy of this conformation comprises the loss of resonance energy plus residual repulsion energy. This non-planar form is more stable than the planar form because the loss of resonance energy is more than offset by the decrease in non-bonded repulsion. The ease of twisting of the above type depends on the variation of resonance energy with angle of twist (θ). The resonance energy varies as $\cos^2 \theta$ (Dewar[95]) and clearly resonance remains considerable even for quite large angles of twist. Steric effects leading to a decrease of resonance in a molecule have been recognized for some time and are referred to specifically as the 'steric inhibition of resonance'.

Aliphatic molecules having more than two double bonds can have more than two planar geometries. Thus, the compound $CH_2=CH—CH=CH—CH=CH_2$ can exist as the forms *Figure 2.24a, b, c* and *d*, for example, the geometries *Figures 2.24a* and *b* and the pair, *Figures 2.24c* and *d*, are interconvertible by rotation about a classical single bond (partial double bond) and are conformations. The structures, *Figures 2.24a* and *c*, are only interconvertible by a rotation about a double bond and so represent

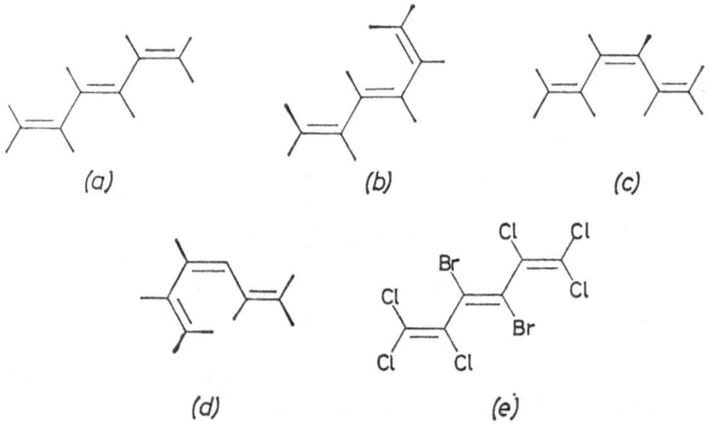

(a) *(b)* *(c)*

(d) *(e)*

Figure 2.24

distinct geometries. However, the central double bond about which the interconversion *Figure 2.24a → Figure 2.24c* takes place is part of a conjugated system and has a bond order <2; this reduces the overlap barrier height although, as is generally true, this is insufficient to make *Figures 2.24a* and *c* conformations. The most stable structure, as with but-1,3-diene, is that in which all the double bonds are staggered (*Figure 2.24a*). This staggered form is essentially planar but the analogous form, *Figure 2.24e*, has large halogen substituents and steric effects result in a non-planar molecule.

Aromatic molecules—Resonance stabilizes the planar forms of a conjugated system and creates an overlap barrier to their interconversion. When more than two planar geometries for a conjugated system can be drawn some of these will involve interconversion by rotation about double bonds (order <2) and some involve rotation about single bonds (order >1). Rotations about double bonds represent distinct geometries and rotations about single bonds represent conformations. Further, the steric factor may operate in some or all of these geometries causing deviation from planarity; non-planarity is easily caused because of the $\cos^2 \theta$ resonance energy relationship. These principles, involving a steric and a resonance barrier, may be illustrated for aromatic systems.

Acetophenone (*Figure 2.25a*), styrene (*Figure 2.25b*) and chlorobenzene (*Figure 2.25c*) exist in only one planar form:

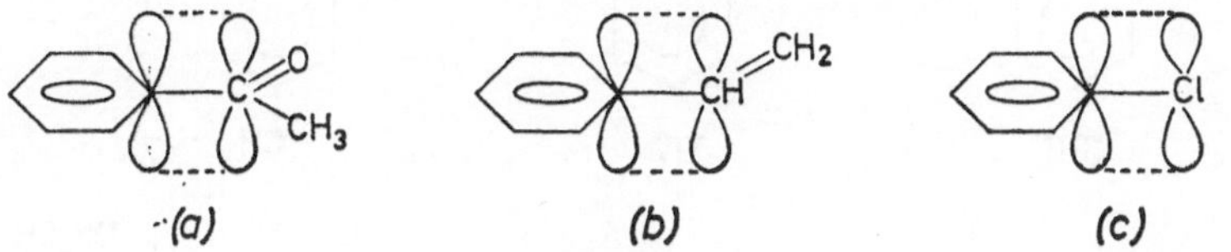

Figure 2.25

no steric factor operates and the planar form is the preferred conformation. The molecule of 2-methylacetophenone has two planar forms (V) and (VI); in (V) a considerable $CH_3 \cdots CH_3$ interaction occurs while in (VI) only a slight interaction between the

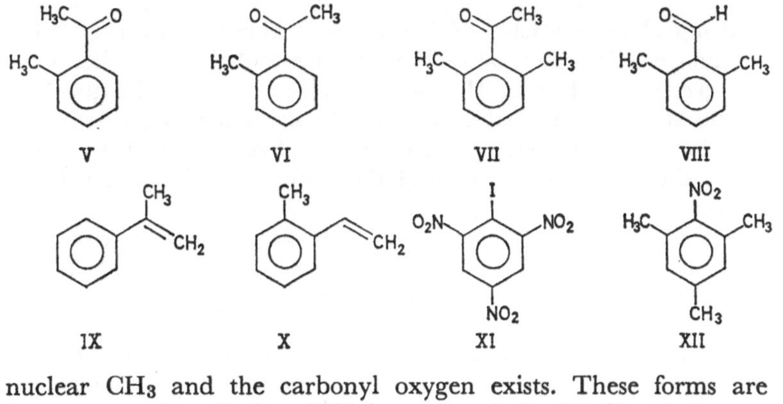

nuclear CH_3 and the carbonyl oxygen exists. These forms are interconverted easily but (VI) is preferred. In 2,6-dimethylacetophenone only one planar form is possible; the strong $CH_3 \cdots CH_3$ interaction occurs and the resultant most stable form is non-planar. The corresponding aldehyde experiences little repulsion and one planar form (VIII) exists. The styrenes (IX) and (X) are similarly non-planar; picryl iodide (XI) has its two ortho nitro groups rotated out of the nuclear plane but the para nitro group is in this plane; the nitro group in (XII) is also non-planar with the

71

F

ring. The molecules (XIII)–(XVIII) also illustrate the steric inhibition of resonance. Conjugation of the carbonyl group and the ring is prevented in (XVI) when n = 5–9. The non-planarity of (XVII) is brought about by rotation about the ring —CO bond

XIII   XIV   XV

XVI   XVII   XVIII

and not about the CO—N bond; this occurs because ring —CO resonance is less than CO—N resonance and the latter bond has the greater bond order (Edwards and Meacock[96]). The geometry of the alicyclic ring in (XVIII) results in rotation about the ring —O bond and so interfering with ring —O resonance, it is claimed to be the most extreme case of the steric inhibition of resonance (Arnett and Wu[97]).

The steric factor operates also in more complex aromatic systems, thus, the conjugated diphenyl molecule (XIX) can adopt one planar form and steric interaction of hydrogens, in the ortho positions shown, results in the actual molecule being non-planar, the angle of twist about the central single bond is determined by the point of equilibrium between the decrease of non-bonded repulsion and the loss of resonance energy. In spite of the small interaction between the hydrogens the angle of twist is large (45 ± 10 degrees) as expected from the $\cos^2 \theta$ relationship. Thus, at a twist of 45 degrees there is still 50 per cent resonance between the rings. The diphenyl molecule illustrates a further matter of general stereochemical importance, thus, in diphenyl and most of the other molecules considered the steric barrier is not so high as to prevent rotation at ordinary temperatures. However, the hypothetical rigid planar diphenyl molecule would have a non-bonded repulsion of more than 70 kcal/mole and clearly rotation through such a barrier is impossible at ordinary temperatures. The rotation becomes possible because the C—H bonds are easily bent back; the energy

required for this is much less than the decrease of non-bonded repulsion due to the increased separation of hydrogens.

The total strain (angle strain + residual repulsion strain) in the planar state is, therefore, reduced to about 5 kcal/mole. In diphenyl itself rotation through the steric barrier is possible at ordinary temperatures but the introduction of appropriate substituents, as in (XX) can prevent this.

The compounds stilbene and azobenzene can exist in the pairs of planar forms (XXI)–(XXII) and (XXIII)–(XXIV) respectively. A steric factor does not operate in (XXI) and (XXIII) therefore the molecules are planar. However in (XXII) and (XXIV) repulsion occurs between the hydrogens shown on the rings; this results in the rings of (XXIV) being rotated by 56 degrees, in the same direction, out of the planar position. A similar interaction of rings occurs in (XXV) and they assume a propeller-like arrangement by rotation about the $\overset{+}{C}$—ring bond.

In summary one concludes that (*a*) rotation about a C—C bond is free but a steric factor may operate and create an energy barrier to rotation. A high barrier results in separate geometries, a low barrier gives conformations; (*b*) a single bond with partial double bond character has an overlap (resonance) barrier to rotation. Rotation is easy but a steric factor may operate to hinder rotation more or even to make it impossible at room temperature; (*c*) a double bond with single bond character has an overlap or resonance barrier, the bond remains essentially double in most cases and separate geometries exist; (*d*) the C=C has a strong overlap barrier and separate geometries exist; (*e*) a triple bond with partial double bond character remains essentially a triple bond and rotation is free; (*f*) the C≡C allows free rotation.

In the molecules discussed previously some relief of steric interaction may be had by rotations about classical single bonds or by twists about classical double bonds. However, an important group of aromatic molecules is known in which steric repulsions can be reduced only by the distortion of bond angles or of rings. Thus, the chlorine atoms in 1,2-dichlorobenzene interact and push each other out of the aromatic plane by the bending of bonds. The steric barrier created is low and the atoms can easily slip past each other

XXVI   XXVII   XXVIII

by a vibration at ordinary temperatures. This effect is not apparent in hexamethylbenzene, which is planar, due to the greater elasticity of the alkyl groups. However, the octamethylnaphthalene molecule is non-planar because the 1,8- and 4,5-methyl groups are close and interact strongly; these groups are forced out of the aromatic plane by 0·73 Å. Steric interactions between other methyl groups occur with the result that all the methyl groups lie above or below the plane of the nucleus (Donaldson and Robertson[98]). These internal distortions are called 'intramolecular overcrowding' effects (Bell and Waring[99]). Intramolecular overcrowding effects of more general stereochemical interest are encountered amongst polycyclic compounds. Thus in, the molecules (XXVI) (Herbstein and Schmidt[100]), (XXVII) (Robertson[101]) and (XXVIII) (Newman[102]) certain carbon atoms, such as C_1 and C_{12} in (XXVI), approach to less than the van der Waals contact of 3 Å; high repulsion occurs and the molecule is made non planar. The distortions of these molecules are rather complex but they involve bond lengths and bond angles and are distributed over all the rings so as to minimize the effect on the conjugation in any one ring. Steric barriers in these molecules are frequently high enough to prevent free vibration at room tempera ture.

The above discussion of the steric factor has invoked only repulsive effects, however steric attraction effects can occur also. Thus, the substance 1,2-dichloroethylene can exist as the stable geometries *Figures 2.26a* and *b*. The form shown in *Figure 2.26a* has the chlorine

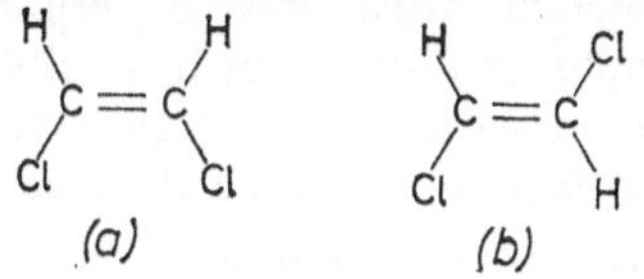

Figure 2.26

atoms at about van der Waals contact and optimum attraction exists. This steric attraction reverses the usual order of stability found in substituted ethylenes and makes *Figure 2.26a*, with the substituents closest, the most stable form by 0·6 kcal/mole.

3

THE FACTORS AFFECTING ATOMIC AND MOLECULAR GEOMETRY: (III) THE ELECTROSTATIC, ENERGETIC AND ENVIRONMENT FACTORS

THE ELECTROSTATIC FACTOR

INTRODUCTION

Electrostatic charges may develop in molecules in three main ways namely (a) due to an atom being ionized, e.g., $R_4\overset{+}{N}$; (b) due to a dative link, for example, $R_3\overset{+}{N}\rightarrow\overset{-}{O}$ and (c) due to atoms of different electronegativity being joined by a covalent bond. The existence of charges in molecules can result in the occurrence of intramolecular and intermolecular attractions or repulsions. These forces will favour a particular conformation in flexible molecules or lead to a development of strain energy in rigid systems. These types of electrostatic interactions may be called the electrostatic factor.

TYPES OF CHARGE DEVELOPMENT IN MOLECULES

Covalent bonds between dissimilar atoms have a partial ionic character; they may be represented as $\overset{\delta\oplus}{A}$—$\overset{\delta\ominus}{B}$ and such a system is called a 'dipole'. The charges in a dipole are regarded as resident at the nuclear centres and so are separated by a distance equal to the bond length. Dipoles originate in bonds due to a property of atoms called 'electronegativity'; this concept, in general form, has

Table 3.1. Electronegativity Values for Atoms

H	2·1	P	2·1	Br	2·8
C	2·5	S	2·5	I	2·5
N	3·0	F	4·0		
O	3·5	Cl	3·0		

been in chemical use for some time (Flurscheim[103]; Lewis[104]; Stieglitz[105]) but a more precise interpretation of it is recent (Pauling[13c], Mulliken[106]; Gordy[107]). The electronegativity of an atom represents its capacity to attract electrons and some important values are given in Table 3.1 (Pauling).

On the simple Lewis model the static electron pair is drawn closer to the more electronegative atom and becomes unequally shared. In molecular orbital terms the electronegativity differences of the bonded atoms results in a modification of the statistics of electron motion so that an electron pair spends most of its time between the nuclei but closer to the electronegative atom. This leads to a 'fattening' of the localized molecular orbital at its more electronegative end. In resonance terms a bond dipole originates from the hybridization of covalent and ionic canonical states; the contribution from an ionic structure becomes important because electronegativity differences stabilize such a form. The polarity of a bond is proportional to the electronegativity difference of the bonded atoms (Malone[108]) and Table 3.1 permits the deduction of the direction and relative degree of polarization of the bonds in any molecule thus, the $\overset{\delta\oplus}{C}$—$\overset{\delta\ominus}{Cl}$ and $\overset{\delta\ominus}{O}$—$\overset{\delta\oplus}{Cl}$ bonds are polarized as shown and the charges are of about the same magnitude.

The values of electronegativities in Table 3.1 apply to saturated atoms and derive from systems in which no other effects operate. Electronegativity actually depends on the atomic state (sp^3C (2.5) $<sp^2C<spC$) and on electronic interactions such as resonance clearly the data of Table 3.1 is limited in its use. Thus, in vinyl chloride, $CH_2 = CH$—Cl, the contribution from the state $\bar{C}H_2$—$CH=\overset{+}{C}l$ gives a net positivity to the halogen atom. Dipolar molecules have an increased cohesive capacity due to the attractive interaction of dipoles in different molecules. This interaction is a second type of van der Waals interaction and is specifically called 'Keesom interaction'. This Keesom[109] interaction may be intermolecular or intramolecular but only the latter is of stereochemical importance to the present discussion. The magnitude of interaction of two dipolar molecules depends on the polarity of the bonds and

Table 3.2. The Dipole Moments of Some Bonds in Debye Units

C—H	0·3D	C—C	0	C=N	0·9	C—S	0·95	C—Br	1·48
N—H	1·31	C=C	0	C≡N	3·6	C=S	2·8	C—I	1·29
O—H	1·53	C≡C	0	C—O	0·85	C—F	1·51	N→O	3·2
S—H	0·68	C—N	0·4	C=O	2·4	C—Cl	1·56	S→O	2·5

their relative orientation. In order to calculate Keesom interaction the polarity of a covalency is stated in terms of the bond dipole moment (bond length × pole strength); the orientation of dipoles in a molecule follows from a knowledge of bond lengths and angles.

The moments of some important bonds are given in Table 3.2; the directions of these moments follow from the electronegativity values of the atoms. The bond moments of Table 3.2 refer to isolated systems in which no other effects occur; as for electronegativities they are affected greatly by resonance.

The interaction of a system of charges, say two dipoles, is not difficult to imagine in a simple qualitative form. However, the mathematical treatment shows the interactions to be rather subtle in nature. Consider two point charges e_1 and e_2 at distances z_1 and z_2 from an origin O (*Figure 3.1*); the electrostatic field potential

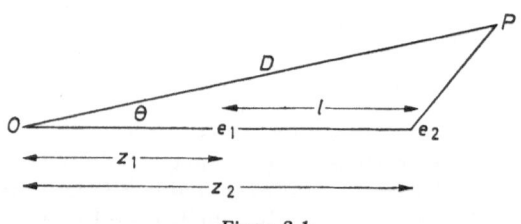

Figure 3.1

at any point P is given by equation 3.1, the derivation of this equation

$$\text{Potential} = \frac{e_1 + e_2}{D} + (e_1 z_1 + e_2 z_2) \frac{\cos \theta}{D^2} +$$

$$(e_1 z_1^2 + e_2 z_2^2) \frac{3 \cos^2 \theta - 1}{2D^3} + \quad . \quad . \quad . \quad . \quad (3.1)$$

assumes that $D > z_1$ and z_2. The term $(e_1 + e_2)$ is called the zeroth moment, the term $(e_1 z_1 + e_2 z_2)$ is called the dipole moment, the term $(e_1 z_1^2 + e_2 z_2^2)$ is called the quadrupole moment, etc. That the term $(e_1 z_1 + e_2 z_2)$ is a dipole moment is seen if $e_1 = -e$ and $e_2 = e$ since then $(e_1 z_1 + e_2 z_2) = e(z_2 - z_1) = el$. Equation 3.1 shows that the potential falls off in the order dipole > quadrupole > etc.; the effective ranges of the potential fields are in this order. It will be clear that a charge or a system of charges in the vicinity of the system e_1 and e_2 will experience forces due to the various dipole, quadrupole, etc., potentials at the point; these forces will be in the above order of the various potentials. An important consequence of equation 3.1 is that molecules such as CO_2 which have two equal and opposite dipole moments and so a resultant of zero do in fact have a quadrupole moment; such molecules can interact and have a cohesive capacity in consequence.

Permanent polarity also exists in ions and in molecules containing co-ordinate links; the electrostatic effects of these forms of polarity are, of course, much greater than those arising from dipoles created in covalent bonds. The operation of these various electrostatic effects for various types of system can be discussed now.

ALIPHATIC MOLECULES

The C—H bond has a small moment $(0.3\ D)$ and therefore in eclipsed ethane the charged hydrogen atoms will repel. Further, a quadrupole repulsion occurs and both of these electrostatic effects reinforce the steric barrier so stabilizing the staggered form (Lassettre and Dean[110]). The inclusion of the electrostatic factor goes some way to accounting for the discrepancy between the calculated steric barrier and the experimental barrier found in ethane.

Halogenated ethanes contain quite strong dipoles (C—Br = $1.48\ D$) and these molecules have been studied closely in order to determine the relative importance of the steric and the electrostatic factors to the rotational barrier heights. A comparison of the steric and electrostatic contributions may be had from a comparison of the potential energy curves for n-butane and ethylene dibromide. These molecules are identical except for the strong dipole factor of the halide since the bromine atom and the methyl group have the same van der Waals radii. Actually the lengths of the C—CH$_3$ and C—Br bonds differ being $1.54\ \text{Å}$ and $1.91\ \text{Å}$ respectively; this difference can be ignored since it has little effect on the degree of overlap.

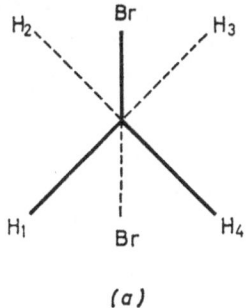

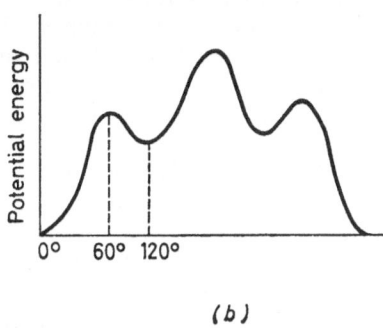

(a)

(b)

Figure 3.2

When a CH$_2$Br group in staggered ethylene dibromide (*Figure 3.2a*), is rotated the potential energy of the molecule increases (*Figure 3.2b*) due to the increase of steric and electrostatic repulsion. The steric repulsion develops principally between Br · · · · H$_1$ and

Br $\cdots\cdot$ H$_2$ but a smaller interaction develops between H$_3$ $\cdots\cdot$ H$_4$. After a relatively small rotation interaction between the two bromine atoms becomes apparent; in ethylene dichloride it has been calculated (Gwinn and Pitzer[111]) that the chlorine atoms are in van der Waals contact after a 50 degree rotation from the staggered position. Thus, in ethylene dibromide, since bromine is larger than chlorine, the potential energy after a 50 degree rotation (say) is due to a large steric Br $\cdots\cdot$ H$_1$ and Br $\cdots\cdot$ H$_2$ interaction, a smaller Br $\cdots\cdot$ Br interaction, small H$_3$ $\cdots\cdot$ H$_4$ interactions and the Br $\cdots\cdot$ Br electrostatic interaction; at a 60 degree rotation the steric interactions Br $\cdots\cdot$ H$_1$, Br $\cdots\cdot$ H$_2$, and H$_3$ $\cdots\cdot$ H$_4$ are at a maximum and a potential maximum is reached. Further rotation rapidly decreases these interactions and the decrease more than offsets the continued increase in the steric and electrostatic Br $\cdots\cdot$ Br interactions. Thus, a potential minimum is found at about 120 degrees rotation and this is the skew form.

The energy differences between the staggered and skew conformations of n-butane and ethylene dibromide are 0·8 kcal/mole and 1·4 kcal/mole respectively. Thus, since in both molecules the steric factor operating in the skew form is the same the electrostatic factor must contribute $1\cdot4 - 0\cdot8 = 0\cdot6$ kcal/mole to the energy of skew ethylene dibromide. Thus when the ethylene dibromide molecule is rotated from the potential maximum eclipsed configuration at 60 degree rotation to the skew form at 120 degree rotation the steric interaction decreases to 0·8 kcal/mole while the electrostatic potential increases to the value of 0·6 kcal/mole. Therefore, in the eclipsed form the steric factor contributes rather more than 0·8 kcal/mole and the electrostatic factor has a value rather less than 0·6 kcal/mole, it follows that the hindering potential to the rotational interconversion of staggered and skew forms is largely steric. A second important general conclusion is that in the preferred skew form the potential energy is about equally contributed to by the steric and electrostatic factors. The heights of the staggered-skew barriers in ethylene dichloride and dibromide are 4·1 and 6·1 kcal/mole respectively.

A rotation beyond 120 degrees in ethylene dibromide increases the potential energy (*Figure 3.2b*); a rigid ethylene dibromide molecule cannot attain the fully eclipsed form since van der Waals penetration of the bromines would result in an infinitely great potential energy. Thus, complete rotation would appear to be impossible. However, as pointed out previously the distortion of valency angles requires little energy and easily occurs under the influence of repulsive forces. Thus, as the halogen atoms approach the repulsions widen the CĈBr and a complete rotation becomes possible. The energy difference between staggered and fully eclipsed ethylene dichloride is calculated as 4·5 kcal/mole, it will be slightly larger in ethylene dibromide. In hexachloroethane, where all the chlorines interact in the fully eclipsed form, the energy barrier has been estimated experimentally as 10–15 kcal/mole.

In simple ethylenic systems containing dipoles, such as 1,2-dichloroethylene, the repulsion between the negatively charged chlorines is greatest in the stable form having the halogens closest together. However, the van der Waals attractions offset the dipole repulsions and this form has the greatest overall stability. The diacetyl (I), dimethyl oxalate (II) and chloroacetyl chloride (III)

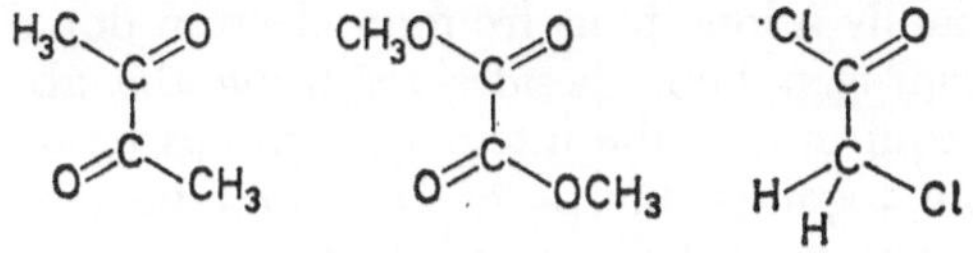

molecules have the preferred conformations shown; the steric and electrostatic factors reinforce each other to stabilize these conformations.

ALICYCLIC MOLECULES
The operation of the steric factor in substituted cyclohexanes favours an equatorial orientation of substituents. However, in a

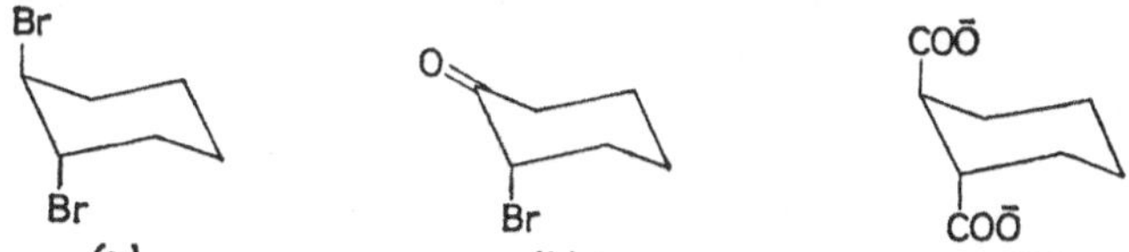

Figure 3.3

polysubstituted cyclohexane, this conclusion must be modified if the substituent introduces a dipole factor. Thus, the molecules of *Figure 3.3* are most stable as axial conformations because the negative charges on the substituents are closer in the equatorial form and the repulsion is greater.

Aromatic Molecules

It will be clear from the examples discussed previously that the electrostatic factor will operate frequently in aromatic molecules; thus, the steric barrier in diphenyls can be augmented when the ortho groups are polar.

The Hydrogen Bond

The previous discussion has considered the stereochemical importance of general dipoles; in certain molecules a rather special interaction involving dipoles can occur so stabilizing a particular geometry, this interaction is called the hydrogen bond. The hydrogen bond derives its name from the fact that a polarized hydrogen atom, $\overset{\delta\oplus}{H}$, is always involved. This polarized species interacts with electrons, usually a lone pair, from an electron donor. The formation of a hydrogen bond besides requiring the above electronic conditions requires that the interacting groups be disposed appropriately close together in space. It should be pointed out that although the nature of the interacting species necessary to hydrogen bonding seems clear the detailed theory of the interaction is still in an unsatisfactory state.

The covalent attachment of the weakly electronegative hydrogen atom to a strongly electronegative atom, such as oxygen or fluorine, strongly polarizes the covalency and the hydrogen atom becomes highly positive. Further, the small van der Waals radius of the normal hydrogen atom is made even smaller when it becomes polarized because the attached electronegative species withdraws electron cloud. A dipole of this type because of the high charge on the hydrogen atom and its low van der Waals radius can approach an electron donor species closely and so interact exceptionally strongly. This hydrogen bonding can be intermolecular or intramolecular but only the latter is important at this point. Hydrogen bonding is found in various molecular types and ortho substituted aromatic molecules are suited particularly to the effect since the substituents are orientated in close proximity. In ethylene chlorohydrin (IV) the $\overset{\delta\oplus}{H}$ of the hydroxyl group and the $\overset{\delta\ominus}{Cl}$ can hydrogen bond when the groups have a skew relation and this stabilizes this form, relative to the staggered form, by 0·95 kcal/mole. The dicarboxylic acid $HOOC \cdot CH = CH \cdot COOH$ can exist as the two geometries (V) and (VI): these structures are separated by a rotational barrier, about the double bond, of 15·8 kcal/mole for the interconversion (V) and (VI). The operation of the steric factor makes (VI) less stable than (V) but this is considerably modified by hydrogen bonding and (VI) is only 6 kcal/mole less stable. The molecules (VII) and (VIII) illustrate the stabilization of enol forms by hydrogen bondings. The cyclohexane (IX) has hydroxyl

$CH_2OH \cdot CH_2Cl$		
IV	V	VI

VII	VIII	IX·	X

XI	XII	XIII	XIV

groups in the 1,3-position; the high repulsion of two axial 1,3-groups is offset by hydrogen bonding. The o-hydroxybenzaldehyde (X) is stabilized as shown, by a hydrogen bonding interaction. The molecule of o-nitrophenol (XI) is of interest since a hydrogen bond might be established with the doubly bonded oxygen or the co-ordinately bonded oxygen; the latter might be expected to be used but in fact resonance in NO_2 group makes the oxygens equivalent in charge and the co-ordinate linked formulation is unreal. In a similar way (XII) and (XIII) are stabilized by hydrogen bonding. The meta- and para-isomers of (X)–(XIII) have no internal hydrogen bond since the interacting groups are too widely separated. The molecule (XIV) is a subtle case of hydrogen bonding, one expects the lone pair on nitrogen to interact with the hydroxyl group, however, the geometry of the cyano group makes it geometrically impossible for such an interaction to take place. Nevertheless, (XIV) does in fact have an intramolecular hydrogen bond; the hydrogen of the hydroxyl group reacts with the π-electrons of the triple bond of the cyano group. That the π-electrons of a multiply bonded group can function as the donated or basic electrons for hydrogen bonding is of recent discovery but a variety of examples of this type are now known (von Schleyer[112]). It is to be noted that by definition hydrogen bonding is always an attractive inter-action while general dipole interaction can be either attractive or repulsive. The stabilization due to hydrogen bonding depends on the particular molecule and is generally between 1·5 and 7·5 kcal/mole.

One concludes that the electrostatic factor is of some importance in determining geometrical fine structure; it may act attractively or repulsively and it may reinforce or oppose the steric factor.

THE ENERGETIC FACTOR
Introduction
It has been pointed out that, at ordinary temperatures, the staggered and skew forms of n-butane are interconverted readily but that the two 1,2-dimethylethylenes are stable under these conditions. It follows that, at ordinary temperatures, the 1,2-dimethylethylenes may be obtained as pure substances while the n-butanes cannot be separated. However, at a sufficiently low temperature n-butane may consist entirely of the staggered form while at a high enough temperature the two forms of 1,2-dimethylethylene are inter-convertible and equilibrate. It is clear that the stereochemistry of a molecule is not a fixed property but depends on the energy of the molecule, i.e. on the energy of the environment. The dependence of molecular geometry on energy may be called the energetic factor.

The Fundamental Thermodynamic and Rate Equations
A system consisting of molecules of geometry G_1 may transform itself spontaneously, under given conditions, into a system in which the molecules have geometry G_2. Classical thermodynamics shows that such a transformation is possible only if there is a decrease in free energy for the change. The reverse change $G_2 \rightarrow G_1$, would result in an increase of free energy and so does not take place under the given conditions. Alternatively the initial system G_1 may provide a final state consisting of an equilibrium mixture $G_1 \rightleftharpoons G_2$. The same thermodynamical principle applies thus the system G_1

spontaneously goes to the equilibrium mixture because the latter system has a lower free energy than the initial system. When an initial system is proceeding to the equilibrium state there is a continuous decrease in free energy until the equilibrium state is established. The equilibrium state may be established also by starting from an initial system G_2 and again a free energy decrease occurs. Thus, the free energy of an equilibrium state is a minimum with respect to either system G_1 or G_2 and each of these latter systems will give the equilibrium state spontaneously.

An equilibrium system is described quantitatively by equation 3.2:

$$K = e^{\frac{-\Delta F}{RT}} = e^{\frac{-\Delta H}{RT}} \cdot e^{\frac{\Delta S}{R}} \quad \quad . \quad . \quad . \quad (3.2)$$

K is the equilibrium constant and its magnitude measures the position of equilibrium, a large value of K for the equilibrium $G_1 \rightleftharpoons G_2$ shows that more of G_2 than G_1 is present. The equation shows that the constant K is related to the difference in the molar free energies of substances G_1 and G_2. If $F_{G_1} > F_{G_2}$ then ΔF is negative and clearly the larger the value of F_{G_1} relative to F_{G_2} the greater is the value of K. The equilibrium constant may be defined in terms of ΔH $(H_{G_2} - H_{G_1})$ and ΔS $(S_{G_2} - S_{G_1})$. An exothermic reaction has ΔH negative while a positive value of ΔH is found for an endothermic reaction. Thus, the larger H_{G_1} relative to H_{G_2} the greater is K; similarly the larger S_{G_2} relative to S_{G_1} the greater is K. The transformation of G_1 into G_2 will be more complete, i.e. the equilibrium mixture will consist largely of G_2, the greater the decrease in heat content (ΔH) and the greater the increase in entropy (ΔS). Equation 3.2 can be simplified in two important ways, firstly, since volume changes are negligible in most transformations ΔH can be equated to the change in energy content (ΔE), secondly entropy changes can be ignored frequently $(\Delta S = 0)$. Thus one can write $K = \exp(-\Delta E/RT)$.

The concept of free energy change defines the possibility of a transformation (or equilibration) of a system occurring, however, even though a decrease in free energy is possible it does not follow that the transformation actually takes place at a reasonable rate. Thus, complete knowledge about a transformation requires both thermodynamic and kinetic information. The theory of absolute reaction rates postulates that the interconversion of two conformations or geometries (G_1 and G_2) proceeds via an intermediate geometry (G^*) called the transition state. The transition state derives from an initial geometry by a process of internal rotation or vibration and corresponds to that intermediate geometry of highest energy. Thus, in the interconversion of the two stable 1,2-dimethylethylenes the perpendicular state is the transition state. The transformation of G_1 to G^* is envisaged as being an equilibrium reaction during the whole course of the overall $G_1 \rightarrow G_2$ transformation. When the transformation leads to a $G_1 \rightleftharpoons G_2$ equilibrium then the final system will contain the equilibria $G_1 \rightleftharpoons G^*$ and $G^* \rightleftharpoons G_2$. It follows from the concept of the transition state that the conversion $G_1 \rightarrow G_2$ or the conversion of G_1 to an equilibrium state is only possible if the molecules G_1 can acquire the energy necessary to convert them into the transition state. The rate of attainment of a final state (G_2) or an equilibrium state ($G_1 \rightleftharpoons G_2$) is determined by the rate of transformation of G_1 into G_2. The formation of the transition state is a relatively slow process but once formed a transition state goes rapidly to another species (G_2); the rate of formation of the transition state obviously determines the rate of the conversion $G_1 \rightarrow G_2$ or the rate of establishing the equilibrium state. The rate of transformation of G_1 into G_2 will involve certain thermodynamic functions of the transition state, this rate is given by equation 3.3.

$$k = \frac{RT}{Nh} \cdot e^{\frac{-\Delta F^*}{RT}} = \frac{RT}{Nh} \cdot e^{\frac{-\Delta H^*}{RT}} \cdot e^{\frac{\Delta S^*}{R}} \quad . \quad . \quad (3.3)$$

The symbol k is the specific rate constant for the transformation $G_1 \rightarrow G_2$ and ΔF^* is the difference in molar free energy of the transition state and the initial substance G_1. The quantities ΔF^*, ΔH^* and ΔS^* are called the free energy, heat and entropy of activation. The magnitude of k will be greater, and reaction faster, the smaller the difference $H^* - H_{G_1}$, since H^* is always greater than H_{G_1}; k is greater the greater the increase in ΔS^*. The simplifications described earlier may be used and $\Delta H^* = \Delta E^*$ (the energy of activation); ΔS^* can be ignored frequently. The simplified equation is $k = (RT/Nh) \cdot e^{\frac{-\Delta E^*}{RT}}$ and the smaller the difference $E^* - E_{G_1}$, i.e. the smaller the activation energy, the faster will be the reaction.

In summary, the stereochemical importance of the thermodynamic factor is that if a molecule can have more than one geometry then the preferred conformation or the most stable geometry is that of lowest free energy. When the entropy change can be ignored the most stable stereochemistry is that of lowest heat or energy content. Entropy cannot be ignored in some cases and then the simplified conclusion does not hold; thus cases are established where the preferred or most stable geometry has the higher heat content, its stability is determined by its higher entropy. When the various geometries of a molecule can be equilibrated then equation 3.2 quantitatively defines the point of equilibrium. The rate factor shows that a transformation or equilibration of geometries is attainable only if the energy to provide the transition state is available. The rate of the change is given by equation 3.3 and is determined by free energy, heat, energy and entropy of activation.

ROTATIONAL INTERCONVERSIONS

Molecular rotation can occur in two general ways; the rotation of the molecule as a whole occurs and rotation of parts of the molecule, about valency bonds (internal rotation), may exist. Internal rotation only is of direct stereochemical importance although not all of the possible internal rotations have stereochemical consequences. Thus, in the ethane molecule only rotation about the C—C bond has a stereochemical effect; a rotation about C—H bonds does not change the geometry. The rotation of the molecule as a whole is important in that it constitutes a form of molecular energy and so contributes to the free energy, heat content etc. of a molecule; all forms of rotational energy are quantized.

The capacity for internal rotation in a molecule depends on the amount of its internal rotational energy. This rotational energy increases with temperature and clearly the restriction of internal rotation is temperature dependent; at a high enough temperature the rotation becomes free. The presence of an energy barrier may prevent the rotation of a group in a molecule and so the group undergoes oscillation about the position of minimum energy. Thus, in ethane, at an appropriate temperature, the methyl groups twist about the staggered position but do not undergo complete rotations in general. Statistical thermodynamics or the equipartition principle shows that the vibrational energy associated with group oscillations is $RT/10^3$ kcal/mole, thus the energy and amplitude of oscillation is temperature dependent as expected. When the temperature is such that the vibrational energy equals or exceeds the energy of the barrier to rotation then free rotation is possible. At a temperature of $T = 300\,°K$ (i.e. ordinary temperature) the value of $RT/10^3$ is 0·6 kcal/mole but in ethane, for example, the rotation barrier is 3 kcal/mole hence the molecules exist with their methyl groups undergoing small oscillations about the staggered position. However, the value 3 kcal/mole is an activation energy which is attained

G

easily at room temperature. Thus, during motion through space, the ethane molecules collide and there is an interchange of energy, because of collision some ethane molecules acquire the necessary activation energy and so rotate. At 1500 °K free rotation is possible and the substance no longer consists largely of staggered forms.

The energy barrier separating the staggered and skew forms of ethylene dichloride is 4·1 kcal/mole and the equipartition equation shows that free interconversion should not occur at room temperature. The heat content difference between the conformations is 1·20 kcal/mole, the skew form having the higher heat content. If entropy can be ignored then $\Delta F = \Delta H = 1·20$ kcal/mole and the skew form has the potentiality for spontaneous transformation into the staggered form, although without considering the rate factor one does not know that it does. The transition state for the staggered-skew change is the eclipsed form and the activation energy is 4·1 kcal/mole, the activation energy for the reverse change is $4·1 - 1·20 = 2·90$ kcal/mole. These are quite low activation energies and although free rotation is disallowed the skew and staggered forms can obtain the required energy by collision processes and so an equilibrium is established between the conformations. The equilibrium constant K is calculated from equation 3.2 by substituting in the value $\Delta F = 1·20$ kcal/mole; it is found that gaseous ethylene dichloride consists of 80 per cent of the staggered form and 20 per cent of the skew form at equilibrium. Higher temperatures facilitate the attainment of equilibrium and increase the percentage of skew form present.

In the case of molecules such as ethylene dichloride the entropy difference between the conformations can be ignored. When a molecule can exist in a variety of different geometries of different degrees of rigidity then the most rigid form has the lowest entropy. The molecule of ethylene chlorohydrin can exist in staggered and skew forms; the skew form has the halogen atom and the hydroxyl group close together and an internal hydrogen bond is established. Thus, due to the hydrogen bonding, the skew form is of lower energy and is more rigid than is the staggered form. The skew form has 950 cal/mole less heat content than has the staggered form but the latter has 3·7 entropy units (e.u.) more entropy. The thermodynamics shows that it is the difference in entropy which determines the overall stability of the conformations of ethylene chlorohydrin and therefore the staggered form has the lower free energy. Thus the transformation skew → staggered form is the spontaneous one and equilibrated gaseous ethylene chlorohydrin contains rather more of the staggered form. More recently, other cases have become known where entropy differences actually outweigh energy content differences and determine the direction of transformations.

The free energy differences between equatorial and axial forms measure the stabilities of these conformations; the equatorial forms of monomethyl, ethyl, i-propyl and t-butylcyclohexane are 1·8, 2·1, 3·3 and 5·4 kcal/mole respectively more stable than the axial forms. The interconversion of axial and equatorial forms takes place via the boat form and this latter structure represents the transition state. The activation energies are attainable easily at ordinary temperatures and the equilibrium mixture contains more of the equatorial form. It will be clear that the activation energy for interconversions in ethane and methycyclohexane is derived from nonbonded interactions; in ethylene dibromide the activation energy is largely steric but there is also an electrostatic contribution.

The overlap barriers in ethylene and stilbene are 60 kcal/mole and 40 kcal/mole respectively. A rotation about the double bond in stilbene gives two different stable geometries but, as is generally true, the difference in free energy is not very great. At ordinary temperatures these activation energies of 60 kcal/mole and 40 kcal/mole cannot be attained and interconversion is impossible. At higher temperatures molecules can acquire the necessary activation energies and equilibration occurs. The stilbene geometry in which the two phenyl groups are distant has the lower free energy and the equilibrium mixture contains more of this form.

The interconversion of conjugated systems may involve high or low activation energies depending on whether rotation occurs about a double bond with partial single bond character or about a single bond with partial double bond character. In conjugated systems the activation energy is largely due to the loss in resonance energy although the steric factor frequently contributes; molecules such as but-1,3-diene involve rotation about a single bond with partial double bond character and the activation energies are always low; equilibria are established easily at room temperature and the system contains more of the form having least non-bonded interaction since this has the lowest free energy.

Vibrational Interconversions

Molecules possess quanta of vibrational energy and this results in atoms and groups of atoms in molecules undergoing vibration; the nature of these vibrations depends on the geometry and complexity of the molecule. A simple, but important mode of molecular vibration is that of the atoms along their valency bonds, thus, the vibrational energy causes atoms to be displaced on either side of their equilibrium separation and so vibration is set up along the direction of the bond. The amplitude of vibration, like the rate of rotation, depends on the energy of the environment but, unlike rotation, molecular vibration cannot be frozen out even at the absolute zero. The residual vibrational energy of a molecule at the absolute zero is called the zero point energy.

The property of vibration along bonds means that the stereochemical descriptions of molecules given previously are only average pictures. Thus, in ethylene molecules, for example, vibration along the C—H bonds occurs and the CH_2 groups undergo torsional oscillation about the C=C, however, these effects are small and ethylene is described justifiably as a planar molecule having characteristic bond lengths. It will be clear that due to vibration, a molecule may have an infinite number of geometries, however, only that geometry corresponding to the equilibrium separation of atoms represents an energy minimum and all others are stereochemically unimportant. However, certain molecules may adopt, by vibration, more than one geometry in which normal bond lengths occur and each of these geometries corresponds to an energy minimum and is important.

In certain molecules some of the more gross vibrational modes are of stereochemical importance since they produce geometries corresponding to energy minima. The ammonia molecule and its derivatives are well established cases of this vibrational effect. Thus, the pyramidal ammonia molecule is envisaged as undergoing vibrational inversion via a planar form (*Figure 3.4*); this vibrational mode was suggested first by Meisenheimer[113] and it has been confirmed since by physical methods (Barker[114], Wall and Glockler[115]).

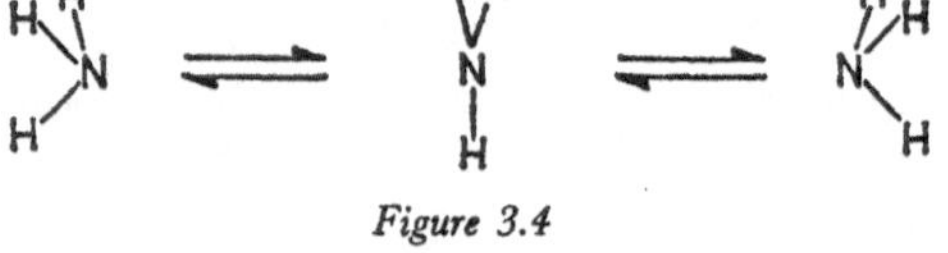

Figure 3.4

The inversion process in ammonia itself is stereochemically unimportant but inversion in amines of the type $R_1R_2R_3N$ gives a different conformation. The nitrogen atom in pyramidal ammonia uses p orbitals having some s character; the planar nitrogen atom uses sp^2 orbitals and the energy required to attain the planar state is a measure of the energy required for the change in hybridization. The planar state is clearly the transition state for the inversion and the activation energy is 6 kcal/mole (Manning[116]); it is angle strain energy. At ordinary temperatures energies of 12–15 kcal/mole are attainable and therefore the inversion occurs easily under these conditions and an equilibrium exists. These considerations show that the two (identical) geometries produced by inversion and corresponding to energy minima are, in fact, conformations. The frequency of oscillation of the ammonia molecule is $2 \cdot 3 \times 10^{10}$/sec (Cleeton and Williams[117]). Substituted ammonia molecules have larger masses attached to the nitrogen atom and this reduces the frequency of oscillation. However, such variations do not suffice, to stabilize the geometries and even at very low temperatures substituted ammonias consist of an equilibrium mixture of the two conformations.

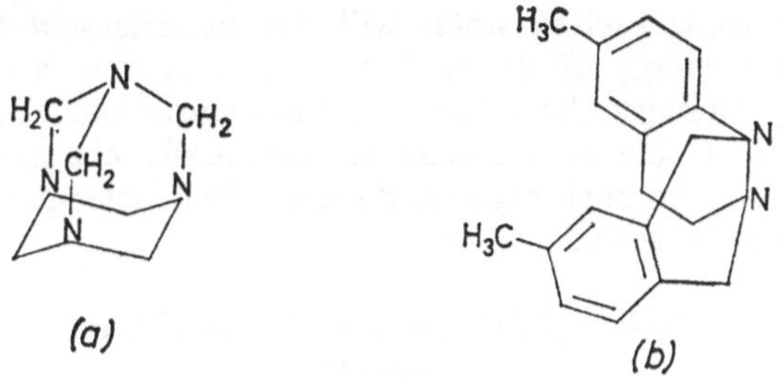

(a)

(b)

Figure 3.5

The stabilization of geometries of aliphatic nitrogen compounds is possible, however, in certain molecules in which the nitrogen atom forms part of a ring, the molecules of hexamine (*Figure 3.5a*) and Tröger's base (*Figure 3.5b*) are examples of rigid systems containing trivalent nitrogen.

The phosphine molecule might be expected to undergo vibrational inversion like ammonia and indeed earlier the frequency of oscillation was given as 5×10^{6}/sec, it has been shown recently (Costain and Sutherland) that in fact the oscillation time for PH_3 is $1 \cdot 4$ years and the two different geometries for species of the type $R_1R_2R_3P$ represent stable molecules and not conformations. The pyramidal carbanion and carbon radical, like ammonia, have a low activation energy of inversion.

The vibrational inversion of ammonia is a particular illustration of the general principle that all non-planar species may be considered, theoretically, to undergo vibrational inversion. Thus, tetrahedral carbon systems were envisaged as capable of vibrational inversion via a planar form (Werner[118]), actually the activation energy of the conversion is so high that the bonds of the molecule would break preferentially but even if this were not so it would take a molecule, such as methane, 10^9 years to switch over the high energy barrier (Hund[119]). The activation energies for species such as $R_3\overset{+}{S}$, R_2SO_2, $R_3\overset{+}{P}\overset{-}{O}$ etc., also prevent vibrational inversion.

TRANSLATIONAL AND ELECTRONIC ENERGIES

The translational and electronic energies of molecules, like rotational and vibrational energies, contribute to the energy content etc., of a system. The quantum levels occupied by electrons in atoms determines the stereochemistry of the atom and this is clearly the atomic state factor. Stereochemistry usually concerns ground state molecules and although the geometry of a molecule is altered by electronic excitation; these excited states will not be discussed here. The translational energy of the molecules of a system is distributed statistically amongst the molecules; the translational energy of a particular molecule in a system is continually changing due to collision, however, such changes do not affect molecular geometry and need not be considered further.

THE ENVIRONMENT FACTOR

INTRODUCTION

The previous discussions of molecular geometry have been concerned, essentially, with molecules in isolation; in practice bulk substances or solutions are employed and it becomes relevant to inquire as to the effect of the state of aggregation on molecular geometry. Any such effects will be intermolecular in origin and may be called the 'environmental factor'. This factor involves two main problems namely the stereochemistry of molecules in different states of aggregation and the effect of temperature on the geometry of molecules in a given state of aggregation. It will be clear that the environment factor depends on the operation and magnitude of intermolecular forces; in bulk substances these forces constitute a net attraction and are responsible for the cohesion of gases, liquids and solids.

TYPES OF INTERMOLECULAR FORCES

Intermolecular forces are described under the general name of van der Waals forces and three main types may be distinguished, namely (a) 'dispersion' or 'London forces', (b) 'Keesom forces' and (c) 'Debye forces'.[120]

London forces operate between all atoms and molecules and their magnitude is proportional to $1/d^7$ where d is the separation of the particles; these forces like all atomic and molecular forces, are electrostatic in origin and arise in the following way. Atoms or molecules having electron clouds which are not symmetrical about nuclei have permanent dipoles, when symmetrical electron clouds are present the atoms or molecules have, over a period of time, zero dipole moment. Thus, in the spherically symmetrical ground state hydrogen atom for example, the electron and the proton constitute a dipole at a given instant. These dipoles may interact electrostatically and the electronic motions will synchronize to produce a disposition of protons and electrons such as $+ - + -$. The thermal motion of molecules, particularly in gases, will tend to oppose the attractive orientation of the dipoles but nevertheless, in bulk substances, there is a net attraction of particles and even species such as neon will experience this cohesive London interaction.

A second type of van der Waals attraction is that originating from dipole-dipole interaction, this is specifically called Keesom interaction and is proportional to d^{-4}. Hydrogen bonding, as will be seen from the examples given, in part at least involves Keesom interaction. Molecules having permanent dipoles can also induce polarization in other molecules in their vicinity and such induced polarization is in a direction so that attraction results. This dipole-induced dipole interaction is called Debye interaction and is proportional to d^{-7}. These three types of van der Waals forces are the main types found but ion-ion, ion-dipole, ion-induced dipole, ion-quadrupole, dipole-quadrupole and quadrupole-quadrupole forces may also operate; these forces are proportional to d^{-2}, d^{-3}, d^{-5}, d^{-4}, d^{-5} and d^{-6} respectively. The magnitude of van der Waals forces depends on the separation of the particles but it also depends on the nature of the particles. In crystalline argon the interaction energy is entirely of the London type and is 2·03 kcal/mole; in crystalline ammonia the London, Keesom and Debye interactions are 3·52, 3·18 and 0·37 kcal/mole respectively. Ice due to hydrogen bonding, has a Keesom energy of 8·69 kcal/mole while the London and Debye energies are 2·15 and 0·46 kcal/mole respectively. Thus, in systems other than those in which hydrogen bonding occurs, the London energy is greatest although in systems such as ammonia in which the molecules have considerable polarity the Keesom interaction is large; Debye interaction is invariably small.

Crystalline aliphatic hydrocarbons and their derivatives have the carbon chain in fully staggered form; the chains are arranged in the crystal in a parallel fashion. This crystal geometry is expected since the staggered form is of lowest free energy and, further, since it is most planar the van der Waals interaction will be optimum because molecules can approach closely. Thus, in crystalline aliphatic molecules the energetic and environment factors reinforce and produce a unique geometry of the molecules. The energetic and environment factors do not always reinforce and the problem may be stated in general terms, thus, if a substance may have molecules of conformation A and B of energy difference ΔE kcal/mole, with A more stable, and if these conformations can give crystal structures A_1 and B_1 whose energy difference, due to intermolecular forces, is ΔE_1 kcal/mole, then if B_1 is more stable than A_1 and if $\Delta E_1 > \Delta E$ crystal structure B_1 results and is made up of the least stable conformations. The crystal structure of an aliphatic compound depends on the temperature, at higher temperatures the staggered chains undergo torsional oscillation and even rotation as a whole about their axis although no crumpling due to internal rotation occurs. On melting, some internal rotation becomes possible and other conformations are present; the ratio of the different conformations depends on their free energy differences and on the temperature. In the vapour state, with the considerable decrease of intermolecular forces due to increased molecular separation some crumpling occurs although the fully staggered form still largely exists. The crumpling or coiling of an aliphatic chain in liquids is the basis of an empirical rule relating to reactivity (Newman[121]). Thus, the coiling in n-butyric acid results in the approach of reagents to the carbonyl group being hindered. In completely fluorinated aliphatic hydrocarbons the fluorine atoms on alternate carbons of the staggered form interact repulsively and the chain becomes twisted, after thirteen carbons a twist of 180 degrees has occurred; this twist is present in the crystal since the crystal forces are too small to enforce the staggered structure.

Molecules having permanent dipoles provide an important illustration of the environment factor. Thus, at a sufficiently high temperature ethylene dichloride gas would consist of molecules having free internal rotation about the carbon-carbon bond, actually due to the hindering potential the angular momentum of the CH_2Cl groups is not the same throughout the rotation. At lower temperatures free rotation ceases and the gas consists of an equilibrium mixture of the skew and staggered forms with the latter predominating. Thus, in the gas the geometry of the molecules is determined by the steric, electrostatic and energetic factors and not by the environment factor. However, in liquid ethylene dichloride considerable dipole-dipole interaction occurs between the polar skew forms since the molecules are relatively close together. This Keesom interaction results in an increase in the skew forms present and the liquid contains about equal amounts of the two conformations. In energetic terms and considering only the Keesom interaction it is clear that the hypothetical condensation of a mole of the gaseous staggered form to a mole of liquid staggered form results in no change in Keesom interactions and so no change in the molar energy. This follows since the C—Cl dipoles in the staggered form are of the same magnitude and are orientated opposite and parallel and therefore the molecule has no resultant dipole moment. The orientation of the dipoles in the skew form does not result in their cancellation and the molecule does have a resultant moment and so can interact with other skew forms. Thus, the condensation of gaseous skew forms results in a decrease in the molar energy. The energy differences between the conformations of ethylene dichloride are ΔE gas = 1·2 kcal/mole, ΔE liquid = 0 kcal/mole; for ethylene dibromide ΔE gas = 1·40 kcal/mole and ΔE liquid = 0·65 kcal/mole; these energy differences determine the point of equilibrium, in the gas the staggered form predominates, in liquid ethylene dibromide the staggered form predominates also but less so than in the gas and in liquid ethylene dichloride the amounts of the two conformations are equal.

Crystalline ethylene dihalides are made up of completely staggered forms since this allows closest approach and the development of strong London forces. As pointed out, intermolecular

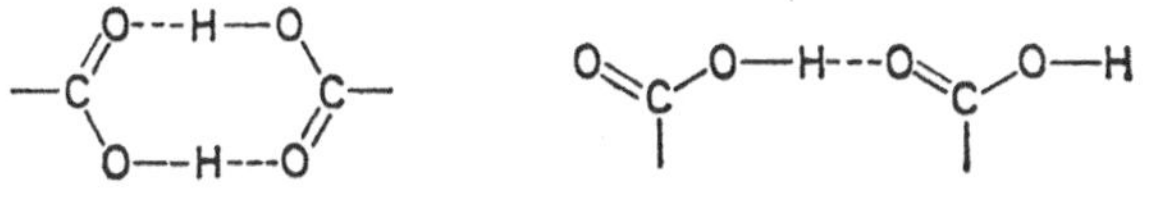

(a)   (b)

Figure 3.6

hydrogen bonding is a form of Keesom interaction. The carboxylic acids are a well-known case of this type and these molecules can interact in the two ways shown in *Figures 3.6a* and *b*. Hydrogen bonding is a strong interaction and is present even in the gaseous state; formic acid has bonding of type *Figure 3.6a* in the gas while in the crystal the interaction is of type *Figure 3.6b*. The dicarboxylic acids such as adipic and glutaric bond by the type *Figure 3.6a* in the crystal; the carbon chain in these molecules has the planar zig-zag form but the optimizing of hydrogen bonding results in slight rotation of the carboxyl groups out of this plane. Special types of geometrical effects arise from hydrogen bonding in molecules such as quinol (Powell and Palin[122]) and urea (Schlenk[123]; Smith[124]) the bonding results in the formation of molecular cages which are capable of trapping other molecules, these molecular cages and enclosures are call 'clathrates'.

'Chair' cyclohexane is more planar than the 'boat' form and the energetic and environment factors reinforce making the crystalline material consist entirely of the 'chair' form; it has been suggested that liquid cyclohexane may contain small amounts of the 'boat' form. The equatorial conformation of substituted cyclohexanes is preferred generally; this conformation is expected to be preferred in terms of the environment factor because it is the more planar form. Gaseous monomethyl- and monochlorocyclohexane are reported as being entirely or almost entirely equatorial and it is expected that the amount of axial form will decrease in the liquid and be non-existent in the solid compound. In the substance 1,2-dichlorocyclohexane the stability of the axial form is increased; a solution of the substance in a non-polar solvent or the substance in the liquid state does contain, in fact, an equilibrium of the biaxial and biequatorial conformations. However, the solid substance is made up entirely of the biequatorial form as expected. In crystalline 1,4-dichlorocyclohexane the halogens are biequatorial but in a non-polar solvent some biaxial form exists although the gas is said to contain only the biequatorial conformation.

Aromatic molecules are generally planar and so are suited

XV   XVI

particularly for close packing in the crystal. However, two complications can exist preventing the parallel arrangement of aromatic systems in the solid. Thus, molecules which can hydrogen bond, such as quinol, give a cage structure and not a parallel array. The second complication arises from the presence of a powerful steric factor. Thus, the planar form of diphenyl is stabilized by the resonance factor but the steric factor favours the non-planar form. In the gaseous state the steric factor is determining and the molecule is non-planar but in the solid, crystal forces enforce planarity. However, in molecules such as 2,2'-dichloro-4,4'-diaminodiphenyl (XV) the aromatic rings are rotated by 36 degrees from the coplanar position in the solid. Intramolecularly overcrowded molecules are similarly non-planar in the crystal.

The environment factor is capable of affecting molecular geometry in a further general way; thus 1,3,5-trioxan (XVI) has $\overset{\frown}{COC} = 110$ degrees in the gas but in the crystal this angle is modified to 105 degrees due to the operation of crystal forces.

One concludes that the environment factor is, in general, a secondary factor, it may produce some modification in geometry and alter the proportions of conformations but rarely has a drastic effect on the expected stereochemistry of the molecule.

4

STEREOISOMERISM

TYPES OF STEREOISOMERISM

INTRODUCTION

The term stereoisomerism (Meyer[125]) describes a property which certain molecules have of existing in more than one, stable, non-superposable, stereochemical form. The phenomenon is, therefore, a direct derivative of molecular geometry and geometrical stability. It was shown earlier that most molecules can exist in an infinite number of different geometries, however, only a few of these are stable enough to merit consideration as stereoisomers. The concept stable, in modern terms, and as applied to stereoisomers, means that the different stereochemical forms of the molecule are detectable though not necessarily isolatable.

In the classical period of stereochemistry two types of stereoisomerism were recognized; these were called *cis-trans* and optical stereoisomerism. The recognition and classification of these two types of isomerism arose from the fact that the permutation of given substituents about either a carbon-carbon double bond or about a tetrahedral carbon atom gives two different molecular geometries for each type of system. This interpretation requires the further assumption that the different geometries are not easily interconvertible and can be isolated as individuals.

More recently, subtle and rather less stable forms of *cis-trans* and optical isomers have been detected by modern physical methods, these isomers are called rotational and vibrational stereoisomers. Although the detection and detailed interpretation of rotational and vibrational isomers is new the possibility that internal molecular rotation and vibration may result in different stereochemistries for a molecule has been recognized for some time (van't Hoff[3c]; Werner[118], Meisenheimer[113]).

The stereoisomers of a molecule are interconvertible in principle and frequently in practice by the absorption of energy. Stereoisomers therefore, are separated by energy barriers and the heights of these barriers are of great practical and theoretical importance. The stereoisomers of a molecule when separated by a high energy barrier may be called primary stereoisomers; a low barrier to inter-conversion defines the stereoisomers as of the secondary type. The high energy barrier makes possible the individual existence of primary stereoisomers at ordinary temperatures whereas secondary isomers are interconverted readily and can exist only as equilibrium mixtures, primary isomers are, of course, the type of stereoisomer encountered in the classical period; the low barrier type requires modern techniques for detection. Although a pure secondary isomer cannot be isolated from an equilibrium mixture a substance may, or may be made to, consist wholly of the most stable secondary form. Thus, a substance may be normally crystalline or may be made crystalline and this frequently results in its molecules assuming one particular stereochemical form.

The mechanism of interconversion of stereoisomers of all types is by either a rotational or a vibrational process. Classical *cis-trans* isomers are separated by high rotational barriers and classical optical isomers have high vibrational barriers. The rotational and vibrational types of *cis-trans* and optical stereoisomers are separated by low energy barriers of the type their name implies; they are classified as secondary stereoisomers. These various types of stereo-isomerism may be discussed now in more detail.

CIS-TRANS ISOMERISM

The molecule of 1,2-dimethylethylene is essentially planar at room temperature because of its π-bonded unit. The characteristic geometry of this unit permits the two permutations, (I) and (II)

| I | II | III | IV |

of the substituents. The molecules (I) and (II) clearly have non-superposable geometries and are, therefore, stereoisomers. A complete analysis shows that the existence of these two stereoisomers is traceable to (*a*) an appropriately substituted parent geometry (in this case a substituted π-bonded unit) and (*b*) to the existence of an energy barrier to the interconversion of the different molecular geometries (in this case a high rotational barrier arising from the rigidity of a double bond). The nomenclature used to describe stereoisomers of the type (I) and (II) is due to Baeyer; it derives from the spatial relations of substituents. Thus, the isomer (I) is called *cis* 1,2 dimethylethylene (*cis* = on this side of; a reference to the methyl groups) and (II) is called the *trans* isomer (*trans* = on the other side of).

A variety of other doubly bonded units permit *cis-trans* isomerism; the $C=N$ and $N=N$ units fulfil both of the above conditions and provide *cis* and *trans* isomers just as does the carbon-carbon double bond. The $C=O$ unit satisfies only the barrier condition since two different permutations of substituents about it are impossible. The triply bonded $C \equiv C$ and $C \equiv N$ units satisfy neither of the conditions. One concludes that appropriately substituted classical double bonds provide primary *cis* and *trans* stereoisomers. This analysis is based on the geometry and rigidity of π-bonds and, therefore, is only true in the absence of any electronic, steric or other factors which may decrease the rigidity of the π-bond or otherwise facilitate the interconversion of the stereoisomers. Thus, the double bond barrier in an allyl carbonium ion (e.g. *cis*—$RCH=CH-\overset{+}{C}H_2$) is reduced in height by conjugation so permitting easy conversion to the *trans* isomer (Jones[126]); the allyl carbanion is a further example (Lanpher[127]). These ions, therefore, in spite of their classical double bond exhibit secondary *cis-trans* isomerism and exist as equilibrium mixtures of stereoisomers.

The internal energy barrier necessary for *cis-trans* isomerism may be created by means other than double bonds and, in appropriate molecules, this may actually lead to *cis-trans* isomers. Thus, a barrier to internal rotation about a σ-bond may be created by a steric factor and a rotational barrier may also result from the resonance stiffening of a classical single bond. A steric barrier separates the staggered and skew forms of ethylene dibromide, for example, and these conformations are designated as *trans* and *cis* isomers respectively. *Cis* and *trans* isomers of this type raise a point of nomenclature since although the staggered form has the true *trans* relationship of the bromine atoms (projected valency angle 180 degrees), the true *cis* form (projected angle 0 degrees) is the non-existent eclipsed conformation. Thus, since the skew form is designated *cis*, a projected angle of 60 degrees and not 0 degrees characterizes *cis* stereoisomers of this type. The molecules of but-1,3-diene and N-methylacetamide illustrate the conjugational barrier; the *cis* and *trans* forms of this latter molecule are shown as (III) and (IV) respectively. The *cis-trans* isomerism associated with a resonance barrier obviously presents no nomenclature problem since projected angles are 0 degrees and 180 degrees for the isomers.

The steric barriers in molecules may be of all heights and depend on the particular molecule. The variations possible in barrier height lead to the existence of both primary and secondary *cis-trans* isomers of this type. Resonance barriers are always low and only secondary isomers of this type are found.

The existence of *cis-trans* isomerism in alicyclic molecules was known classically. This type of isomerism arises naturally from the Baeyer geometries of these systems; thus cyclohexane-1,4-di-carboxylic acid can have the two carboxyl groups disposed in a *cis* (V) or a *trans* (VI) manner. These cyclic *cis-trans* isomers are separated by high vibrational barriers; they are interconvertible, in principle, by the transformation of one or both substituted carbon centres from an essentially tetrahedral into a planar form. This transition state, for example (VII) may then proceed to either

| V | VI | VII |

the *cis* or *trans* isomers. This vibrational isomerization is actually only possible in principle since the energy required to produce the transition state is very large and the molecule would disrupt preferentially.

A system of nomenclature to describe the various types of *cis-trans* isomers is desirable; the previous discussion suggests the following scheme. *Cis-trans* isomerism due to a double bond may be called π-isomerism; it may assume a primary or a secondary form. The steric barrier type of *cis-trans* isomerism may be called σs-isomerism (σ refers to isomerism about a single bond and s refers to the steric cause) the *cis-trans* isomerism originating from conjugation effects may be called σc-isomerism (σ has the previous meaning and c refers to the conjugation factor). Stereoisomers of the σs-type can be primary or secondary but the σc-type are always secondary. Mulliken[128] describes σc-stereoisomers as *s-cis* and *s-trans* the prefix s refers to the single bond. Cyclic *cis-trans* isomers may be called r-isomers (r refers to the ring); these *cis-trans* forms are primary in type.

Historically, the maleic (VIII) and fumaric (IX) acids were the first examples of *cis-trans* isomers; these molecules show primary π-isomerism. The form of stereoisomerism of these molecules was predicted by van't Hoff[36] on the basis of his tetrahedral double bond model for ethylene. The predicted configurations for (VIII) and (IX) were verified by Wislicenus[129] and *cis-trans* isomerism became a reality.

The existence of *cis* and *trans* isomers of many molecules is an experimental fact. However, it is to be realized that the described theoretical deduction on simple geometrical grounds is not a complete analysis. The reality of the isomerism for a particular molecule can only be concluded theoretically by showing that not only are the stereoisomers geometrically possible but also that they are thermodynamically stable. The steric factor usually operates to destabilize a *cis* isomer; the stability of *cis-trans* isomers and the barrier height separating them are related, thus the less stable a *cis* isomer, say, the lower the barrier to its isomerization.

OPTICAL ISOMERISM

The four substituent groups (i.e. CH_3, C_2H_5, H and OH) of the *sec*-butyl alcohol molecule can be permutated about the central tetrahedral atom in the two different ways (X) and (XI). These

two geometries are non-superposable and therefore represent stereoisomers. These isomers differ in a special way from the *cis-trans* type in that one isomer is the mirror image of the other. This relationship is easily seen by imagining the broken line between (X) and (XI) to represent a plane mirror. Molecules having non-superposable mirror images are defined geometrically as dissymmetric and the type of stereoisomerism shown by such dissymmetric molecules is called optical stereoisomerism. The mirror image isomers, such as (X) and (XI), are called enantiomers and they may be interconverted, at least in principle, by a vibrational mechanism and via the planar transition state (XII). This interconversion process requires a change of the central carbon atom orbitals from sp^3 to dsp^2 hybrids and for this process a high activation energy is needed (Kincad and Henriques[130]). This energy quantity is much greater than that required for the scission of a carbon-carbon bond and therefore the vibrational mechanism, though a useful concept, is a hypothetical one. One concludes that optical isomers exist because of (*a*) a dissymmetric molecular geometry (in this case arising from the appropriate substitution of a tetrahedral centre) and (*b*) due to the existence of an energy barrier to interconversion (in this case a high vibrational barrier). Enantiomers have identical energies but again their existence cannot be predicted solely by the geometrical criterion; they must be thermodynamically stable also.

The dissymmetry condition for optical stereoisomerism is satisfied by a wide variety of molecules which may be based on atoms other than carbon. The important general types of dissymmetric molecules are, $R_1R_2R_3R_4C$ (sp^3 carbon types), $R_1R_2R_3R_4N$ (sp^3 nitrogen), $R_1R_2R_3R_4\overset{+}{N}$ (sp^3 nitrogen), $R_1R_2R_3R_4P$ (sp^3 phosphorus), $R_1R_2R_3R_4\overset{+}{P}$ (sp^3 phosphorus), $R_1R_2R_3S^+$ (sp^3 sulphur), $R_1R_2R_3S$ (sp^3 sulphur), $R_1R_2R_3N$ (p^3 nitrogen), and $R_1R_2R_3P$ (p^3 phosphorus). These types of molecules may easily be shown to be dissymmetric by drawing or constructing their models and comparing with their mirror images. The geometry of ethylenic and acetylenic molecules and of carbonium ions does not permit dissymmetry. The pyramidal carbon radical $R_1R_2R_3C\cdot$ and the tetrahedral carbanion $R_1R_2R_3C^-$, however, are dissymmetric species. The oxygen and sulphur atom p^2 states and the halogen atom p state cannot provide dissymmetric molecules.

These general types of dissymmetric molecules may take either an aliphatic or alicyclic form; the barriers to their isomerization are vibrational in all cases. The optical stereoisomers based on sp^3 hybrid bonds are separated by high energy barriers and molecules of this type, therefore, show primary isomerism. Stereoisomers from p^3 nitrogen, carbanions and radicals have low vibrational barriers and exist only in secondary form at ordinary temperatures but p^3 phosphorus compounds show primary isomerism.

The combination of molecular dissymmetry and an energy barrier, necessary for optical stereoisomerism, can arise in other

H

ways. Thus, a steric barrier to rotation about a σ-bond can exist and if the two molecular geometries on either side of the barrier are dissymmetric then they represent optical stereoisomers. This case of isomerism is not to be confused with the similar situation which gives *cis-trans* isomers. Thus, for optical stereoisomerism both molecular geometries are dissymmetric whereas for *cis-trans* isomers at least one geometry is symmetrical and usually they both are. The molecule (XIII) is an example of a steric barrier optical isomer;

the non-bonded interaction between the ortho substituents results in the inhibition of rotation of the rings about the single bond joining the nuclei and also in the formation of a non-coplanar molecule as shown. The non-coplanar molecule is dissymmetric even though it does not contain specific centres of the type described previously. These rotational barriers being of steric origin may be of all heights depending on the particular molecule and, therefore, both primary and secondary isomers of this type exist.

Intramolecular overcrowding introduces dissymmetry into ethylenic or aromatic molecules by distorting their normal planarity. The structure (XIV) is an overcrowded molecule and its enantiomers are interconverted by a vibrational process. The steric nature of overcrowding will result in all degrees of enantiomer stability and both primary and secondary isomers are found.

A miscellaneous group of dissymmetric molecules different to the above types exists; examples are the spiran (XV) and the allene (XVI). The molecules in this group are based on π-bonds (XVI) or on sp^3 carbon (XV) and, therefore, exist only as primary stereoisomers.

Optical stereoisomers containing specific centres may be called c-optical isomers (c refers to the presence of a specific centre). Steric rotation barrier isomers may be described as os-optical isomers; the symbols have the significance described previously. Optical isomerism due to overcrowding may be called i-isomerism (i refers to the intramolecular effect). The inherent dissymmetry of molecules of the miscellaneous group is responsible for their stereoisomerism and they may be called g-isomers (g indicates that specific geometrical characteristics are involved). A distinction may be made in the origin of dissymmetry in the above types of stereoisomers. Thus, the dissymmetry of c isomers is associated with the presence of a specific centre whereas os, i and g isomers have no such centres; the dissymmetry of this latter type of molecule is called general molecular dissymmetry.

The resonance factor can create *cis-trans* isomers but it can oppose only and not create optical stereoisomers. Thus, the operation of resonance in molecules enforces planarity and planar molecules are symmetrical hence cannot exist in enantiomeric forms.

ROTATIONAL ISOMERISM

This form of isomerism has been implied in the previous discussions; it results from the hindering of rotation about classical single bonds in a molecule. Rotational isomers, by definition, are separated by low energy barriers and so are necessarily secondary. This criterion is essential if confusion is to be avoided in designating stereoisomerism which arises from the hindrance of rotation about single bonds. Thus, steric effects hindering rotation may produce both primary and secondary isomers of the *cis-trans* and optical types. However, of these types only the secondary isomers are described as rotational isomers. *Cis-trans* isomers separated by a resonance barrier are necessarily rotational isomers. The stereoisomers (III) and (IV) of N-methylacetamide are examples of rotational *cis-trans* isomers; a diphenyl of the type (XIII) but having ortho groups which create only a low barrier is a case of rotational optical stereoisomerism. A more complex case of rotational stereoisomerism is that of ethylene dibromide; this molecule has more than one internal steric energy barrier. The existence of more than one internal barrier means that the number of stable conformations is increased beyond two. The ethylene dibromide molecule may exist as the symmetrical staggered or *trans* form and as the two dissymmetric skew or *cis* conformations, further, the two skew forms are enantiomers. Thus rotational isomerism may lead to both optical and *cis-trans* types of isomerism being shown by one molecular species. A further example is 2,3-dibromobutane; this molecule contains centres of the type $R_1R_2R_3R_4C$ and so necessarily shows optical stereoisomerism. The enantiomers of this compound are derived, of course, from sp^3 carbon atoms and are quite stable in themselves, however, they can exist as equilibrium mixtures of rotational forms. Thus, (XVII), (XVIII) and (XIX) are the 'star' diagrams representing the rotational forms of an enantiomer of the

compound. These stereoisomers differ from those of ethylene dibromide in that they are all optical stereoisomers, nevertheless it is valuable to be able to specify the rotational forms in terms of the spatial relationships of the substituents. Thus, the methyl groups of (XVII) are transoid whereas in (XIX) the bromine atoms are transoid and the methyl groups are cisoid. In order to define the rotational forms of the optical stereoisomers of a compound such as 2,3-dibromobutane it is necessary to choose reference substituents; the isomers (XVII) and (XIX) are defined as cisoid and transoid respectively by reference to the bromine atoms. The terms cisoid and transoid are necessary for molecules of the above type since these molecules contain specific centres and are, therefore, cases of optical isomerism and not *cis-trans* isomerism.

Rotational isomers have been detected only recently although van't Hoff recognized the possibility of their existence in his principle of free rotation. Modern theory goes far beyond that of van't Hoff; it introduces the concept of energy barrier in detailed form and is able to specify the possible stable conformations.

VIBRATIONAL ISOMERISM

This form of isomerism is analogous to rotational isomerism. Vibrational isomers are secondary and examples of vibrational optical isomers have been encountered previously in the case of molecules based on p^3 nitrogen, carbon radicals and carbanions. Vibrational. *cis-trans* isomers are known also.

Historically, the principle of vibrational isomerization, which is the counterpart of the principle of free rotation, was introduced by Werner in order to explain the interconversion of enantiomers of sp^3 carbon. A more correct application of the vibrational concept is due to Meisenheimer[113], thus it was suggested that enantiomers of the particular type $R_1R_2R_3N$ could not be isolated because they are easily interconverted vibrationally. The Meisenheimer hypothesis has been proved correct as discussed previously in connection with the ammonia problem.

TABLE 4.1

Gross nature of isomerism	Type		Form	Example
Cis-trans	I	π-	a Primary (classical)	Substituted ethylenes
			b Secondary, rotational (non-classical)	Substituted allyl carbonium ion
	II	σs-	a Primary (classical)	Terphenyls
			b Secondary, rotational (non-classical)	Substituted ethanes
	III	σc-	a Secondary, rotational (non-classical)	Substituted amides
	IV	r-	a Primary (classical)	Substituted alicyclics
Optical	I	c-	a Primary (classical)	Substituted saturated aliphatics
			b Secondary, vibrational (non-classical)	Substituted ammonias
	II	σs-	a Primary (classical)	Substituted diphenyls
			b Secondary, rotational (non-classical)	Substituted diphenyls
	III	i-	a Primary (classical)	Substituted polycyclics
			b Secondary, vibrational (non-classical)	Substituted polycyclics
	IV	g-	a Primary (classical)	Allenes

Rotational and vibrational stereoisomers differ in a fundamental respect from primary isomers in that the environment factor considerably affects the composition of an equilibrium mixture made up of rotational or vibrational stereoisomers. Thus, the secondary stereoisomer number for a compound is environment dependent; the primary isomer number is essentially independent of this factor.

The geometry of a molecule can be changed by electronic excitation as well as by rotational and vibrational motions. However, ones discussion of stereochemistry and stereoisomerism is limited to molecules in their electronic ground states and therefore these excited state geometries will not be considered further. The above introductory discussions of the types of stereoisomerism may be elaborated now. The various types of stereoisomers are summarized in Table 4.1; the description of a stereoisomer as classical does not necessarily mean that the particular stereoisomer type was isolated actually during the classical period but that it has been obtained by the methods of classical stereochemistry though perhaps more recently. Non-classical stereoisomers have been detected only by the application of more recently developed techniques.

5

OPTICAL STEREOISOMERISM:
(I) GENERAL THEORY

INTRODUCTION

The existence of the primary enantiomers of *sec*-butyl alcohol follows naturally from the tetrahedral geometry of sp^3 carbon and a high activation energy for their isomerization. Historically, the isolation of primary optical stereoisomers led to the introduction of geometrical ideas (the tetrahedral carbon atom) to interpret them whereas *cis-trans* isomerism was predicted.

Pasteur[131] obtained two tartaric acids and Wislicenus[132] two lactic acids which were found to show a subtle form of isomerism. Thus, these pairs of molecules were shown to be made up of the same structural units and to have identical chemical and physical properties with the important exception that one isomer of a pair rotates plane polarized light positively while the other rotates it to an equal extent negatively. The two tartaric acids have specific rotations in aqueous solutions of $[\alpha]_D^{20} \pm 11\cdot98$ degrees and aqueous solutions of the lactic acids have specific rotations of $[\alpha]_D^{15} \pm 3\cdot82$ degrees; the term optical stereoisomerism derives, from this optical property of the isomers. It was recognized very early that the optical activity of stereoisomers persists in all states of aggregation and this led Pasteur to suggest that the form of isomerism they show is due to dissymmetry at the molecular level. At this time (1848) classical valency theory had not assumed a final form and so molecular dissymmetry could not be interpreted in structural terms. Some years later two theories, of somewhat different form, were introduced by van't Hoff[3a] (September, 1874) and by Le Bel[4] (November, 1874) to account for molecular dissymmetry and optical stereoisomerism.

The theory of Le Bel, influenced by the thinking of Pasteur, focused attention on the geometry of the molecule as a whole and considered the symmetry or dissymmetry of the molecule to be the primary property. Thus, the molecules CH_4 and CCl_4 do not exist as optical isomers and so are symmetrical. Le Bel was willing to allow a tetrahedral distribution of hydrogens about the carbon centre for CH_4 and a square planar distribution of chlorines for CCl_4 because both arrangements give symmetrical molecules. Thus, Le Bel treated molecular geometry as a problem analogous to that of crystal form rather than associating it with the characteristic geometrical properties of atoms. The valency theory of Kekulé influenced van't Hoff and he added to this theory the concept that the valencies of atoms had characteristic distributions in space. On the basis of this theory he approached stereoisomerism by building up molecules from their individual atoms. Actually, van't Hoff restricted the theory to the carbon atom and apparently envisaged this as a tetrahedrally shaped piece of matter having substituents attached to the apices by the undefined valency forces. This concept of a specific geometry for the carbon atom and so for molecules was van't Hoff's real theoretical contribution. The van't Hoff model is very similar to the quantum mechanical sp^3 carbon atom since sp^3 orbitals occupy considerable space. However, the idea of a tetrahedron of matter diverged somewhat from the Kekulé-Couper method of representing molecules using the all-important line bond. A simplified model, which has been in general use for some time and which is closer to the classical representation, uses the atoms symbol and depicts the valencies as four tetrahedrally radiating lines.

The theories of Le Bel and van't Hoff allowed that the carbon atoms in some molecules need not have their valencies distributed in a regular tetrahedral manner. This was an easy matter for Le Bel since for an appropriate molecule, such as *sec*-butyl alcohol, an irregular tetrahedral model has the fundamental property of dissymmetry just as has a regular tetrahedral model. Van't Hoff considered that only when the four substituents attached to a carbon centre are identical is their tetrahedral distribution a regular one; different substituents produce different forces of interaction and so irregularity. On the van't Hoff theory most optical isomers have the irregular tetrahedral shape, a conclusion which, as shown earlier, may be justified in quantum mechanical terms. Irregularities in the tetrahedral valency distribution of an aliphatic carbon atom are slight and therefore the angle between valencies having a regular tetrahedral distribution is an important general stereochemical datum; it is called the tetrahedral angle and its value is 109 degrees 28 minutes. It was clearly recognized by van't Hoff that the geometrical relationship of enantiomers accounts for their identical physical properties of melting point, solubility, free energy etc.

The equal free energies of enantiomers contrasts with the free energy differences between *cis* and *trans* isomers.

THE CRITERIA FOR MOLECULAR DISSYMMETRY

A detection of molecular dissymmetry constitutes a detection of optical stereoisomerism. A knowledge of the barrier height separating the stereoisomers informs as to whether the stereoisomerism is primary or secondary in type. A molecule which is superposable on its mirror image is symmetrical, if it is non-superposable then it is dissymmetric. A testing for the superposability of mirror images is obviously a means of detecting molecular dissymmetry. This method of symmetry analysis requires the representation, on paper, of the three dimensional geometry of molecules and their mirror images and subsequently the mental manipulation of the drawings in order to test for superposability. An alternative procedure is to construct and compare models of molecules and their mirror images. These methods are cumbersome, although they are useful sometimes, and generally the simpler method of examining a three dimensional diagram of the molecule for the elements of symmetry is used. The significance and detection of the elements of symmetry a molecule may have, can now be discussed.

THE PLANE OF SYMMETRY (σ)

A plane of symmetry is one which divides a molecule into halves so that each half is the mirror image of the other. The molecules of water (I), benzene (II) and methane (III) have a plane as

I   II   III

shown. The plane of symmetry in a tetrahedral molecule, for example (III), passes between two hydrogen atoms and cuts the carbon atom and the other two hydrogen atoms into halves; it will be noted that this requires the reasonable assumption that the atoms cut are symmetrical with respect to the plane as is so if the atoms are assumed to be spherical. Other planes of symmetry are possible, of course, in molecules of the above types. All planar molecules have at least one plane of symmetry but, of course, it does not follow that non-planar molecules, for example (III) have no planes of symmetry.

THE CENTRE OF SYMMETRY (i)

If, within a molecule, a point exists so that lines through the point encounter exactly the same environment on each side of the point then the point is a centre of symmetry. The molecules of benzene

IV   V   VI

(IV), staggered ethylene dibromide (V) and chair cyclohexane (VI) have the i's shown.

THE FOURFOLD ALTERNATING AXIS OF SYMMETRY

The stereochemically important symmetry elements a molecule may have are more fundamentally described in terms of the so-called alternating axis. A molecule has an n-fold alternating axis if a rotation of $360/n$ degrees about the chosen axis followed by a reflection in a plane perpendicular to that axis brings the molecule into a position indistinguishable from its original one. It is easily seen that a molecule having a onefold or a twofold axis has a plane or a centre of symmetry respectively. In stereochemistry the plane and centre of symmetry are conceptually simpler than the alternating axis and also have specific uses in stereochemical discussion. However, other alternating axes may be present and a complete symmetry analysis requires their detection; the most important other alternating axis is the fourfold type.

A molecule having none of these three elements of symmetry is dissymmetric; such molecules have non-superposable mirror images and so show optical stereoisomerism. In practice the two simple criteria namely the absence of a plane and a centre of symmetry suffice to define most molecules as dissymmetric. However, some molecules may have neither a plane nor a centre of symmetry and yet possess a fourfold axis so making it superposable on its morror image. This possibility was pointed out by Mohr[133] and Aschan[134] for the molecule (VII) and an example (IX) has been synthesized recently (McCasland, Horvath and Roth[135]). The molecule (VII) has a tetrahedral carbon centre carrying enantiomeric substituents. The molecule has no centre of symmetry and clearly has no plane of symmetry since such a plane must cut two substituents and because these are dissymmetric the plane cannot cut them into

VII   VIII

IX

mirror image halves. The tetrahedral molecule (VII) is arranged with the A^+ groups in the plane of the paper while one A^- projects forward from this plane and the other A^- projects behind this plane. A rotation of (VII) through an angle of 360/4 degrees about the axis a is easily seen to give (VIII); a reflection of (VIII) in the plane p gives (VII) and therefore the axis a is a fourfold alternating axis. In the molecule (IX) the groups R are $\pm$ menthyl; the molecule has no centre or plane of symmetry but has no optical isomers because of the fourfold axis.

In summary, a molecule is capable of optical stereisomerism if (a) it is not superposable on its mirror image or (b) it has no alternating axis present. In practice, it is easiest to test for alternating axes and it usually suffices to show that a plane or centre of symmetry is not present. The analysis of a molecule for the symmetry elements solves the problem of its optical stereoisomerism. However, this is a general mathematical procedure and may be simplified for the particular group of dissymmetric molecules which contain the so-called asymmetric centres.

THE DETECTION OF CENTRES OF ASYMMETRY

As indicated previously, the dissymmetry of the *sec*-butyl alcohol molecule is traceable to (a) the tetrahedral sp^3 carbon atom and

(b) to the attachment of four different groups to this centre. Atoms, such as this sp^3 carbon, which due to their geometry and their particular substitution, create a dissymmetric molecule are called asymmetric centres (van't Hoff). Molecules containing asymmetric centres constitute by far the largest group of dissymmetric molecules and of these the majority have carbon or nitrogen centres. A variety of types of asymmetric centre is known but the quadricovalent tetrahedral centre is the most common.

Asymmetric carbon (or other) atoms have four (or other number) different substituents attached; such centres are detected simply by inspecting the Kekulé-Couper structure of the molecule: their presence implies optical stereoisomerism. The detection of asymmetric centres clearly constitutes a simplified symmetry analysis since it requires no drawing or manipulation of molecular geometry. The method applies to molecules of any degree of complexity whether or not their overall geometry is known.

The number of asymmetric centres a molecule has present determines the number of stereoisomers possible; the actual stereoisomer number is dependent also on whether the centres are similar or different. The carbon centres (carrying COOH) in the tartaric

```
                CHO        CHO        CHO        CHO           CHO
                 |          |          |          |             |
  COOH          CHOH       CHOH       CHOH ──────►CHOH    +    CHOH
   |             |          |          |          |             |
  CHOH          CHOH       CHOH        R₁         CHOH          CHOH
   |             |          |          |          |             |
  CHOH          CHOH       CHOH       XIII        R₂            R₂
   |             |          |                     XV            XVI
  COOH          CHOH      CH₂OH
    X            |         XII
               CH₂OH
                XI                    CHO        CHO           CHO
                                       |          |             |
                                      CHOH ──────►CHOH    +    CHOH
                                       |          |             |
                                       R₁        CHOH          CHOH
                                                  |             |
                                      XIV         R₂            R₂
                                                 XVII          XVIII
```

acid molecule(X) are similar asymmetric centres since they carry the same four substituents but the four centres of the hexose molecule (XI) are different ones. The derivation, from first principles, of the total number of stereoisomers for a substance may be illustrated by (XII). The molecule is drawn in the form (XIII) and as such contains only one asymmetric centre; the two enantiomers may be represented non-geometrically as (XIII) and (XIV). A second centre is now allowed into each of these enantiomers; each one gives a pair of stereoisomers namely (XV)–(XVI) and (XVII)–(XVIII). The introduction of a further centre into each of the four isomers (XV)–(XVIII) completes (XII); each of the isomers gives a pair making a total of eight. The eight stereoisomers may be paired off as enantiomers; thus (XV) and (XVIII) may constitute a pair of enantiomers but not (XV) and (XVI); this correlation will be considered later.

The number of isomers a compound can have may be stated in the form of a simple rule; thus if n different asymmetric centres are present in a molecule then 2^n optical stereoisomers exist and are made up of $2^n/2$ pairs of enantiomers. This rule fails if any of the centres are similar and the number of stereoisomers is always less than 2^n. Tartaric acid has three ($<2^2$) stereoisomers but sixteen (2^4) hexoses exist as eight pairs of enantiomers.

THE DETECTION OF GENERAL MOLECULAR DISSYMMETRY

Dissymmetric molecules fall into two groups, namely those having asymmetric centres and those having what is called general molecular dissymmetry. This latter group of molecules depend for dissymmetry on an inherent molecular geometry which is not traceable to the presence of asymmetric centres. There is no simplified symmetry analysis possible for these molecules; the actual (or hypothetical) geometry of the molecule must be examined for the absence of symmetry elements. The detection of optical stereoisomerism, therefore, in molecules having general molecular dissymmetry requires a detailed knowledge of molecular geometry. In view of the detailed geometrical knowledge required it seems difficult to understand how classical stereochemistry and stereoisomerism were able to develop successfully.

In the classical period the various assigned molecular geometries were entirely hypothetical and although some of these, for example van't Hoff's tetrahedral carbon atom, have proved correct, others such as Kekulé benzene or the Baeyer alicyclic models have proved to be incorrect or approximations to the actual geometry. Thus, Kekulé benzene has no centre of symmetry but quantum mechanical benzene has. In spite of these incorrect models, which lead to an incorrect symmetry analysis, the classical theory very successfully predicted or interpreted the experimentally determined stereoisomer number for aliphatic, alicyclic and aromatic substances. The ability of classical theory to determine correctly the stereo-isomer number was clearly fortuitous; the important matter of the reasons for its success will be considered in detail later.

Modern work has provided much information about actual molecular geometry for many substances, this enables a correct analysis of their symmetry and stereoisomerism to be made. On the classical theory only primary stereoisomers could be considered but with modern knowledge it is possible to determine both primary and secondary isomers.

THE FURTHER ANALYSIS OF MOLECULAR DISSYMMETRY

THE NATURE OF GROUPS ATTACHED TO AN ASYMMETRIC CENTRE

The Isotope Effect

An appropriate atomic state functions as an asymmetric centre when different groups are attached to it; the previous discussion has implied that 'different' simply means constitutionally different and, of course, such differences are readily detected by an inspection of the Kekulé-Couper structure. Isotopic atoms obviously are definable as different because molecules such as R_1R_2HDC can exist as enantiomers. This much sought effect has been proven only recently and the molecule $CHDC_6H_5CH_3$ has been obtained in optically active form $(\alpha)_D^{25} \pm 0.3$ degrees (Eliel[136]). A variety of other deuterium containing optical isomers also has been obtained (Alexander[137, 138]).

The Group Geometry Effect

The criterion that substituents attached to a centre are different if they are constitutionally different happens to suffice as a means of detecting asymmetric centres in most molecules encountered. This simple rule is quite valid but insufficiently comprehensive since in some cases centres carrying constitutionally identical groups can be asymmetric centres. The real meaning of group difference in connection with asymmetric centres is not constitutional but a geometrical one. Thus, a tetrahedral (or other, e.g. p^3 N) centre is asymmetric only if the four (or other number) substituents are not symmetrical about a plane symmetrically cutting the assumed spherical centre. This conclusion follows because for a tetrahedral (or other) system each of the potential planes of symmetry must pass centrally between a pair of substituents and cut the centre and the remaining pair (or other number) of substituents. In view of these facts it is best to detect an asymmetric centre indirectly, thus, if a tetrahedral system has (a) a pair of mirror image substituents and (b) has the remaining pair of substituents symmetrical then a plane of symmetry may be drawn through the tetrahedral centre and the centre is not an asymmetric one. These two conditions must be satisfied simultaneously otherwise no plane of symmetry through the centre exists and the centre is asymmetric.

Asymmetric centres are to be regarded as units of asymmetry and even for some monocentred systems their presence does not necessarily reflect the symmetry properties of the molecule as a whole. The asymmetric centre uses only the plane of symmetry criterion as a test of its asymmetry and while the test is quite valid for identifying such centres it must be restricted to the particular centres. Thus, the planes employed to test a centre must pass through the centre but, when the molecule as a whole is considered, other planes can exist and these may be planes of symmetry. A molecule which is symmetrical as a whole is not symmetrical about a plane passing through an asymmetric centre in the molecule.

Substituents fall into four classes namely (a) those of different constitution and different (non-superposable) geometry (b) those of different constitution and the same (superposable) geometry (c) those of the same constitution and the same geometry (d) those of the same constitution and different geometry. Substituents of class (a) or (b) clearly cannot be mirror images and four such groups make a tetrahedral centre asymmetric. This is the type of asymmetric centre usually encountered and the non-mirror image relationship of constitutionally different groups is the fundamental basis of the simple method of identifying such centres by inspection. The isotope effect is clearly an example of substituents of class (a).

Substituents of class (c) may or may not be mirror images and also they may or may not be symmetrical. Groups, such as Br, are, of course, both mirror images and symmetrical and cannot make a centre asymmetric. However, this type of substituent may derive from an enantiomer and then, of course, they are neither mirror images nor symmetrical. Type (d) substituents clearly can derive only from stereoisomeric parents but again they may or may not be mirror images and also they may or may not be symmetrical. Thus, if two substituents derive from a pair of enantiomers they are mirror images but are not symmetrical while if a substituent derives from a *cis* isomer and another from the *trans* isomer clearly they are not mirror images but are symmetrical.

Substituents deriving from optical or *cis-trans* stereoisomers provide what may be called the 'group geometry effect'; this effect results in certain centres being asymmetric although this cannot be detected by an inspection of the Kekulé-Couper structure; examples of this effect may now be discussed.

Due to asymmetric substituents—As pointed out, asymmetric substituents derived from the same parent enantiomer have the same constitution and geometry but are neither mirror images nor symmetrical. A centre having four such groups, as in (XIX), is

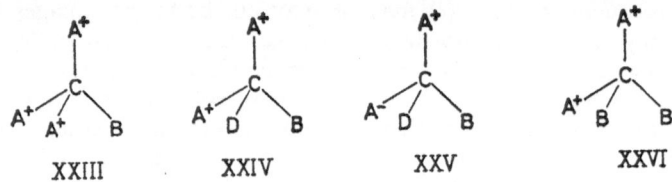

therefore an asymmetric centre, when a centre is substituted by groups of both class (*c*) and (*d*), as in (XX), (XXI) and (XXII), a pair of groups are mirror images but the other pair are not symmetrical (since they are enantiomers) and so the centre is asymmetric. The molecules (XX) and (XXI) are dissymmetric but (XXII) is symmetrical, in spite of the presence of an asymmetric centre, because of the presence of a fourfold alternating axis. It is to be realized that (XXII) is only one stereoisomer of a group represented by the same Kekulé-Couper structure; (XIX)–(XXI) are some other examples of stereoisomers of this group.

Combinations of substituents of classes (*a*), (*b*), (*c*) and (*d*) are of interest; the molecules (XXIII)–(XXVI) are examples. The

four groups of (XXIII) are not mirror images and although B is symmetrical the groups A^+ are not hence the centre is asymmetric; (XXIV) does not have a pair of mirror image groups but the pair B and D are symmetrical groups hence the centre is asymmetric. The centre in (XXV) is not asymmetric since the molecule has a pair of mirror image groups and a pair of symmetrical groups;

(XXVI) has only a pair of mirror image groups (which are also symmetrical) and so the centre is asymmetric.

The compound trihydroxyglutaric acid is a classic example of a substance which provides stereoisomers which have (or have not) asymmetric centres due to the group geometry effect. The terminal carbon atoms carrying carboxyl groups in this molecule (XXVII) are, by inspection, asymmetric centres. The central carbon atom appears not to be asymmetric at first sight, since it has two constitutionally identical groups. The permutation of substituents

about the central carbon gives the molecules (XXVIII)–(XXXI); the centre in (XXVIII) and (XXIX) is clearly asymmetric but that in (XXX) and (XXXI) is not asymmetric. The carbon centre in (XXX) and (XXXI) is frequently called pseudo-asymmetric but in terms of the fundamental definition of centre of asymmetry the term is clearly in error. The dissymmetric molecules (XXVIII) and (XXIX) give the enantiomers (XXXII) and (XXXIII) respectively, however, these latter are, respectively, identical to (XXIX) and (XXVIII) so that only the two different and enantiomeric forms (XXVIII) and (XXIX) are real. The symmetrical forms (XXX) and (XXXI) complete the total of four stereoisomers.

Due to cis *and* trans *substituents—Cis* (C) and *trans* (T) substituents may act as groups of type (c) and (d). However, they differ from substituents derived from optical stereoisomers in that substituents of type (a) or (b) must also be present in order to make a centre asymmetric. Thus, the molecules (XXXIIIa) and (XXXIIIb) do not have the central atom asymmetric but of the four isomers

119

I

of (XXXIV) those drawn as (XXXVII) and (XXXVIII) are dissymmetric due to the asymmetry of the central carbon atom.

C
C — | C
    C   C

XXXIIIa

C
C / | T
    T   T

XXXIIIb

C(CH₃) = CHCH₃
|
CHOH
|
C(CH₃) = CHCH₃

XXXIV

C
H / | \
  OH   C

XXXV

T
H / | \
  OH   T

XXXVI

C
H / | \
  OH   T

XXXVII

C
HO / | \
   H   T

XXXVIII

Due to rotational and vibrational substituents—(a) Due to secondary stereoisomeric substituents—Substituents which derive from primary optical or *cis-trans* isomers may make a centre truly asymmetric as in (XXVIII) and (XXXVII). A new case arises when groups are derived from secondary isomers; such groups are constitutionally identical and geometrically different but are geometrically interconvertible. Thus, the molecule (XXXIX) has two constitutionally identical cyclohexyl groups which may adopt the secondary stereoisomeric 'chair' (*trans*) or 'boat' (*cis*) forms. A molecule having two 'chair' or two 'boat' forms is symmetrical but if the two substituents are a 'boat' and a 'chair' form then the molecule is dissymmetric and

C_6H_{11}
|
CHOH
|
C_6H_{11}

XXXIX

A⁺
A⁺ / | \
  A⁺   A⁺

XL

A⁺
A⁻ / | \
  A⁻   A⁺

XLI

C
B / | \
  D   T

XLII

C
B / | \
  D   C

XLIII

O—H
B / | \
  D—O
    H

XLIV

exists in enantiomeric form. The dicyclohexylcarbinol is, of course, an equilibrium mixture of these four stereoisomers and the isomer having two chair groups will predominate in the bulk substance. The enantiomers have identical energies and are present in equal amounts; the other isomers are symmetrical and so the equilibrium mixture is optically inactive as found experimentally.

The dynamic interconversion of an enantiomer and, say, the stereoisomer having two 'chair' forms as substituents clearly involves the disappearance and the generation of an asymmetric centre. Centres in molecules capable of being rendered asymmetric as a consequence of the internal rotations or vibrations of substituents may be called quasi-asymmetric centres. The types of molecule which may contain a quasi-asymmetric centre will be clear from previous discussion; molecules of types (XL)–(XLI) and (XLII)–(XLIII) are some examples when the substituents are secondary isomers. It will be clear that if, by some means, the geometry of substituents attached to a quasi-asymmetric centre can be stabilized in their different forms then a true asymmetric centre would be created.

(b) *Due to compound substituents*—Internal rotations or vibrations may result in identical compound substituents being orientated so that they are non-mirror images and so an asymmetric centre can be created. Thus, the constitutionally and geometrically identical hydroxyl groups of (XLIV) may orientate themselves as shown so that the central carbon atom is quasi-asymmetric. The molecule becomes symmetrical when the hydroxyl groups are rotationally orientated as mirror images. The geometrical randomness of substituents in molecules of this type is somewhat controlled by conformational factors and in (XLIV) a symmetrical molecule is probably the most stable one. This form of quasi-asymmetry obviously occurs widely, the exceptions are molecules having simple substituents, such as halogens, which may be regarded as spherical.

The molecules discussed previously, such as (XXXV), which do not have asymmetric centres will have quasi-asymmetric centres since they have compound substituents. A quasi-asymmetric centre due to compound substituents can be converted into a true asymmetric centre by stabilizing the substituents in non-mirror image orientations.

DETAILED MOLECULAR GEOMETRY AND DISSYMMETRY

Classical stereochemistry and stereoisomerism were based on (a) the tetrahedral carbon atom, for aliphatic molecules; (b) the regular planar carbon ring structure for alicyclic molecules; (c) the regular planar hexagonal ring structure for aromatic benzenoid derivatives. The validity of these models was in their success; they correctly predicted or explained the experimentally determined primary stereoisomer number for many compounds. However, even when the fundamental concept, i.e. the tetrahedral carbon, was correct the classical theory was a crude approximation since no knowledge of the possible conformations of compound groups attached to such a centre was available. The existence of quasi-asymmetric centres illustrates this and clearly shows that a fundamental analysis of the symmetry properties and stereoisomerism of a molecule requires a detailed knowledge of molecular geometry. The need for a detailed geometrical knowledge of molecules was appreciated even in the classical period and, as will be shown later, this deficiency was mitigated somewhat by the introduction of the principle of free rotation.

The molecule of ethane, for example, can be constructed in an infinite number of different ways by the joining of two tetrahedral units; thus an infinite number of stereoisomers is possible. The conformations (XLV) and (XLVI) represent two possible conformations and stereoisomers. The conclusion that an infinite

XLV   XLVI   XLVII   XLVIII

number of stereoisomers is possible is clearly incompatible with the fact that ethane actually exists as a 'unique' substance and is incapable of (primary) stereoisomerism. This kind of problem was recognized in the classical period and two solutions were possible, firstly, if the ethane molecule adopts a specific, rigid conformation then since no isomers exist this conformation must be a symmetrical one such as (XLV); the conformation (XLVI) is dissymmetric. The absence of stereoisomers was explained in a second way by the concept of free rotation (van't Hoff). This classical principle postulated the rotation of parts of a molecule about single bonds within the molecule. The principle did not imply mechanically continuous rotation nor that the various conformations which rotation can lead to are equally stable, it did imply that if different conformations exist then they are so freely interconvertible as not to be isolatable as individuals. The two (or other) conformations (XLV) and (XLVI) are related to each other by rotation about the carbon-carbon bond and free rotation implies ease of inter-conversion and so explains the uniqueness of ethane. The concept of free rotation in molecules of the ethane type clearly obviates the need of detailed geometrical knowledge and in this way classical theory was successful.

The van't Hoff concept was incorporated into classical stereochemistry as a general principle and, unfortunately, to the effective exclusion of the possibility that molecules may exist in specific, rather rigid conformations. Modern theory shows that, depending on the molecule, both interconvertible and specific geometries may exist. Classical theory, in the form of the principle of free rotation, while recognizing the conformational possibilities for molecules was quite unable to decide which of these possibilities were real and could say nothing about their relative stabilities. Thus, a classical symmetry analysis of ethane is impossible although fortuitously this does not affect the correctness of the conclusion about primary stereoisomer number. The problem is complicated further when compound groups are attached to an ethane carbon skeleton; as shown, such groups can have a variety of conformations but knowledge of these requires detailed geometrical information about bond angles etc. The classical theory was obviously incapable of attempting an analysis of molecules of this type.

The molecule (XLVII) illustrates the particular problem of compound substituents; only a single tetrahedral centre is present and so the problem is not complicated by the existence of different conformations of the carbon skeleton. The classical analysis of this molecule concludes that only one form can exist since no asymmetric centres are present. The application of the principle of free rotation cannot alter this conclusion as to stereoisomer number but does allow the further deduction that the hydroxyl group may be linear or if angular, produces by rotation about the carbon-oxygen bond, a variety of conformations which are easily interconvertible. A last possibility is that the angular hydroxyl group is orientated rigidly but this must be such that the molecule is symmetrical.

Modern theory and physical methods show, in fact, that the hydroxyl group is angular; conformational theory shows that the symmetrical form (XLVII) is the only stable conformation. However, conformations are not rigid and torsional oscillations of the hydroxyl group about the stable position bring the hydrogen atom out of the plane of symmetry and so make the molecule dissymmetric. It should be pointed out perhaps that a molecular geometry produced by an out of plane oscillation does not represent a stereoisomer in the proper sense because such conformations are unstable in the particular sense that they do not correspond to energy minima. A molecule (XLVII) is optically active during its periods of dissymmetry however, the bulk substance shows no optical activity. Thus, if a molecule of the bulk substance is dissymmetric, at any time, due to the hydrogen atom being rotated $+\theta$ degrees, out of the plane of symmetry then for statistical and energy reasons another molecule in the bulk substance having a rotation of $-\theta$ degrees, is equally probable. These two conformations are 'enantiomeric' and therefore their optical activities cancel each other. In the bulk substance all degrees of out of plane oscillation of the hydrogen atom, up to a maximum of 60 degrees, will exist i.e. $\pm\theta_1$ degrees, $\pm\theta_2$ degrees .... ±60 degrees; these conformations are, of course, different. The compound (XLVII), therefore, is made up of an infinite number of pairs of mirror image conformations together with some symmetrical conformations and, of course, it will be optically inactive. Classical theory was unable to conduct an analysis of the above kind although, fortuitously, it reached the same conclusion as to primary stereoisomer number. The molecule (XLVII) contains a quasi-asymmetric centre in the sense described previously. The molecule (XLVIII) contains an asymmetric centre; the actual conformations of the compound groups in molecules of this type cannot affect the already existing dissymmetry created by the asymmetric centre. This conclusion shows why classical theory was successful in calculating the stereoisomer number for molecules containing asymmetric centres.

In summary, modern theory shows, in a precise way, that molecules may develop quasi-asymmetric centres or quasi-general molecular dissymmetry and these possibilities illustrate the need for detailed geometrical knowledge in order to analyse the symmetry properties and stereoisomerism of molecules. The generation of quasi-asymmetric centres or of quasi-general molecular dissymmetry does not modify, however, the conclusions of classical theory about primary stereoisomer number. This is fortuitous since a steric or other factor might stabilize a dissymmetric form of, say, (XLVII) which would then exist as a pair of primary enantiomers; the steric stabilization of dissymmetric conformations actually is found in the diphenyl series. Molecules of this latter type, fortunately, are rare but had they been encountered in the classical period the development of classical theory in its simple form would have been prevented and progress would have had to await an understanding of steric, electrostatic and other factors. The recognition of the steric basis of dissymmetry in the diphenyls (1926) was, in fact, an innovation in stereochemical theory.

ANALYSIS OF MOLECULAR DISSYMMETRY

The above considerations similarly apply to alicyclic and aromatic molecules; thus the cyclohexane molecule can exist in 'chair' and 'boat' forms, only the 'chair' form has the same symmetry properties as Baeyer's erroneous planar model. Alicyclic molecules having compound substituents may contain quasi-asymmetric centres or have quasi-general molecular dissymmetry. The benzene ring has planes and a centre of symmetry in modern terms but only planes of symmetry in classical terms. However, both theories concur as to primary stereoisomer number for a given aromatic system. Internal rotation problems exist for aromatic molecules having compound groups but only recently has the classical interpretation of these molecules had to be extended by the discovery of aromatic systems having steric factors which can stabilize a dissymmetric conformation.

6

OPTICAL STEREOISOMERISM:

(II) ALIPHATIC MOLECULES HAVING ASYMMETRIC CENTRES

INTRODUCTION

Optical stereoisomerism is a consequence of molecular dissymmetry; the dissymmetry in primary stereoisomers may be due to (a) the presence of asymmetric centres or (b) to general molecular dissymmetry. The type of dissymmetry may be used as a basis for the classification and discussion of primary optical stereoisomerism. Secondary optical stereoisomerism will be discussed as it arises in individual cases. Molecules having asymmetric centres constitute by far the largest group of optical stereoisomers and most frequently the centres are tetracovalent.

MOLECULES HAVING ONE ASYMMETRIC CENTRE

INTRODUCTION

Aliphatic molecules with one asymmetric centre may derive from carbon, nitrogen, phosphorus and sulphur. The molecules (I) (Wislicenus[132]), (II) (Stewart and Allen[139]), (III) (Meisenheimer[140,141]), (IV) (Pope and Peachey[51]), (V) (Meisenheimer[57]), (VI) (Kumli, McEwen and Vanderwerf[142]), (VII) (Pope and Peachey[143]), (VIII) (Harrison, Kenyon and Phillips[144]) are some examples.

The representaions (I)–(VIII) of the three dimensional geometry of molecules, on paper, are called perspective diagrams and for stereochemical purposes these diagrams must be imagined and mentally manipulated in their three dimensional form and, of course, not as planar drawings. The perspective drawings (I)–(VIII) are, of course, drawings of only one of the two possible enantiomers that each compound can have, which particular enantiomer is represented by the drawings is important but of no immediate concern. Thus, the lactic acid enantiomer (I) may or may not represent the geometry of that particular isomer having a dextrorotation of +3·82 degrees. Enantiomers such as (I) are stable and isolatable but the amine (II) shows secondary isomerism and would be 50 per cent isomerized at -180 degrees in about 20 sec (Stewart and Allen). It will be realized that the concepts primary and secondary isomerism refer to isomerization by rotation or vibration and not involving the breaking of bonds. However, certain enantiomers of the general type $R_1R_2R_3R_4N^+I^-$ isomerize at ordinary

temperatures in chloroform solution. This does not mean that these enantiomers should be described as secondary since the isomerization involves bond breakage. Thus, an equilibrium is established $R_1R_2R_3R_4N^+I^- \rightleftharpoons R_1R_2R_3N + R_4I$ and the products of dissociation of the quaternary salt can recombine to form an enantiomer different to the starting one. These conclusions apply also to certain $R_1R_2R_3R_4P^+$ compounds and enantiomers of this type were first obtained as recently as 1947 (Holliman and Mann[56a]). Some cyclic molecules containing p^3 nitrogen have a stereochemistry such that the three valencies are fixed and inversion is impossible; secondary stereoisomerism does not occur in molecules of this type.

As pointed out earlier, the perspective diagrams of molecules are drawings of the actual or hypothesized geometry of molecules.

However, only the essential geometrical features need be shown, thus, the perspective (I) of lactic acid focuses attention on the general tetrahedral structure of the molecule and ignores the detailed geometry of substituents. This form of perspective clearly suffices for the discussion of primary stereoisomerism. In the classical period perspective drawings for one and two centred systems found considerable use, as they do still; however, the perspectives of polycentred systems are more difficult to draw and envisage and these molecules are represented usually by the simpler projection diagram. The projection diagram of a molecule has the advantage that it needs to be envisaged and used only as a planar drawing; these diagrams are drawn and remembered easily and clearly these properties make them a useful means of representing stereoisomers.

The major contribution to classical perspective and projection diagrams is due to Fischer[145] but, as will be discussed later, his perspective diagrams for bi- and polycentred systems are geometrically incorrect ones. Fischer's use of incorrect starting perspectives from which to derive projection diagrams does not invalidate, however, the use of these projections but, of course, they carry with them the defects of the perspectives. Modern theory and physical data have provided the necessary molecular geometries for drawing correct perspective and projection diagrams for bi- and polycentred systems. A perspective diagram derived from an actual molecular geometry may be projected in two important ways, firstly, a Fischer type or modern projection may be derived and secondly, a form of projection due to Newman[146] may be obtained. The classical Fischer and the modern projection types are valuable for aliphatic systems containing any number of asymmetric centres whereas Newman projections are most valuable for bicentred systems. The application of these various types of projection to aliphatic systems will now be discussed.

FISCHER PERSPECTIVES AND PROJECTIONS

The glyceraldehyde molecule (IX) is a monocentred system of some historical importance in sugar stereochemistry; the tetrahedral perspective diagram (X) represents a stereoisomer of this substance. The Fischer projection is derived from this perspective by arranging the latter with its asymmetric centre in the plane of the paper and with two valencies (those carrying H and OH groups) projecting in front of this plane and the other two valencies projecting behind this plane; this specific arrangement of the tetrahedral system is shown in (XI). The groups and valencies of this arranged perspective are now projected on to the plane of the paper to give the Fischer projection diagram (XII) for the particular stereoisomer. The asymmetric centre itself is usually not shown in projection diagrams; it is at the point of intersection of the vertical and the

CHO
|
CHOH
|
CH₂OH

IX

X

XI

CHO
|
H————OH
|
CH₂OH

XII

XIII

CHO
|
HO————CH₂OH
|
H

XIV

CHO
|
HO————H
|
CH₂OH

XIVa

horizontal lines of the diagram. The arrangement (XI) of the perspective (X) for the purpose of projection is actually only one of six possible arrangements of (X), thus an alternative arrangement is (XIII) and the projection of this gives the diagram (XIV). The two projection diagrams (XII) and (XIV) are obviously quite different (non-superposable) and yet represent the same stereoisomer. (Four other different projections are possible.) This difficulty is resolved by the arbitrary selection of one of the six possible perspective arrangements from which to derive a projection diagram; the arrangement (XI) has been selected by convention. The projection of this selected arrangement gives the diagram (XII) and this becomes the unique projection for the particular stereoisomer. It is important to realize that the selection procedure is arbitrary and that no fundamental reason requires the choice of (XI); alternative arrangements could be chosen provided that when a choice has been made all other possible arrangements and their projections are discarded. However, as pointed out, glyceraldehyde has historical importance in sugar stereochemistry and for this reason (XII) is found to be the best form of projection diagram.

A second, enantiomeric glyceraldehyde exists and various projection diagrams can be derived as before, however for purposes of remembering, manipulation and correlation it is obviously best if the projection diagram used for the enantiomer is as similar as possible to (XII). The arranged perspective (XI) can be converted to the enantiomeric perspective simply by interchanging two of the substituents and, for the above reasons, it is best to interchange the hydrogen and the hydroxyl groups; the diagram (XIVa) is the projection used to represent the enantiomer.

The conventionally established specific arrangement procedure described for glyceraldehyde exists only because of the importance of this substance in sugar chemistry. The vast number of other monocentred systems are of no special importance and no specific

$$\text{XV} \qquad \text{XVI} \qquad \text{XVII}$$

$$\text{XVIII} \qquad \text{XIX}$$

arrangement procedure has been adopted although schemes have been proposed (Cahn, Ingold and Prelog[147]). These other systems present no problem provided the arbitrarily chosen perspective arrangement for any particular system is maintained throughout a discussion; thus, the substance $C_6H_5CHOHCH_3$ exists as a pair of enantiomers and these can be represented by the alternative pairs of projections (XV)–(XVI) and (XVII)–(XVIII); a discussion of the stereoisomerism of this substance requires therefore an arbitrary selection of diagram.

MODERN PERSPECTIVES AND PROJECTIONS

In the classical period, during which Fischer devised his projection diagrams, the perspective diagrams used for projection were based on hypothesized molecular geometries. The actual geometry for a large number of molecules is now known and it is from these actual geometries, drawn as perspectives, that modern projection diagrams derive. Modern projections for monocentred systems are, of course, identical to the classical ones since the reality of classical tetrahedral carbon is now established. However, as will be shown later the coincidence applies only to monocentred systems; the modern perspectives and projections for higher centred systems are quite different to Fischer's.

NEWMAN PROJECTIONS

As indicated above modern projections derive from modern perspectives by arranging them in the plane of the paper in the manner described for Fischer perspectives. Newman projections are derived, of course, from modern perspectives but an alternative method of arrangement is used. Thus, for a glyceraldehyde enantiomer the molecule is arranged with the central carbon atom in the paper and with the valency carrying the hydrogen atom (say) projecting behind and normal to this plane. The projection of the groups and valencies of this arrangement on to the plane of the paper then gives the Newman diagram (XIX). Newman diagrams actually find their best use in discussions of bi-centred systems.

The above discussion has considered only the aliphatic sp^3 carbon system but, of course, the method is quite generally applicable to other types of systems. Historically, projection diagrams were devised for specific applications in stereoisomerism, however, they now have a quite general use in stereochemical discussions as was shown earlier in the use of 'star diagrams' (which are Newman projections) for the discussion of the steric relations in ethane.

THE PROPERTIES AND USE OF PLANE PROJECTION DIAGRAMS

The only method discussed, so far, for determining how many optical stereoisomers a compound may have is the classical 2^n rule. The rule only permits the calculation of the primary isomer number (aliphatic or alicyclic) when the n asymmetric centres are different. The use of plane projection diagrams extends this rule by providing a means not only of determining the stereoisomer number for any compound but also of pictorially representing its stereoisomers. The method of plane projections is now a standard method of determining stereoisomer number.

Projection diagrams may be used to determine the stereoisomer number in two ways; these may be called (*a*) the group specification method and (*b*) the general permutation method. These two methods are quite general and may be used with all three types of projection diagram. Thus, in order to determine the stereoisomer number of glyceraldehyde the monocentred Fischer projection diagram outline is first drawn and the groups CHO and CH_2OH then are attached specifically to the vertical of the diagram with the former group uppermost. The stereoisomers then derive by the permutation of the remaining groups about the horizontal of the projection diagram. This procedure gives the two different projection diagrams (XX) and (XXI); they are dissymmetric and enantiomeric and represent the possible stereoisomers of glyceraldehyde.

In using the group specification method it is to be realized that there is no fundamental reason determining the choice of groups for attachment to the vertical, in general, any two groups may be selected without in any way affecting the usefulness of the method. In the case of glyceraldehyde the CHO and CH_2OH groups were, in fact, specifically chosen and arranged but only because the substance has special importance and has the generally accepted and specific form of projection discussed previously. The molecule $C_2H_5 \cdot C \cdot Br \cdot CH_3 \cdot COOH$ is not special; the projection outline is drawn and any two groups say, C_2H_5 and COOH are selected and attached to the vertical in any arbitrary manner. The permutation of the remaining groups about the horizontal gives the two different projection diagrams (XXII) and (XXIII); the substance exists therefore, as a pair of enantiomers.

The second method of determining stereoisomer number is a more general one and obviates the need for a specific selection of groups. However, this general permutation method requires a detailed knowledge of the properties of plane projection diagrams. Fischer's development of projection diagrams was essentially of restricted utilitarian interest; thus he was specifically concerned with the representation of sugar and α-amino acid stereoisomers. This limited application of projection diagrams involved the incidental use of some of their properties but Fischer provided no detailed analysis of projection diagram properties *per se*; these properties may now be discussed.

Fischer projections for enantiomers are themselves mirror images (this property has been invoked earlier); however, it does not follow that mirror image projections always represent enantiomers (this applies only to polycentred systems). A projection diagram cannot be tested for the elements of symmetry in quite the same way as a perspective diagram or a molecular model can be. Thus, a Fischer projection may have a centre of symmetry but this arises as a consequence of projection and is not present in the perspective, i.e. the actual molecule (centres of symmetry can only appear in polycentred projections). A centre of symmetry, therefore, in a Fischer projection is meaningless and the projections are not tested for this element of symmetry. The plane of the paper is necessarily a plane of symmetry for a projection diagram but this symmetry element, like the centre of symmetry, is ignored since it has no meaning in the perspective. However, if a projection diagram for a stereoisomer has one or more planes of symmetry other than that of the paper then the perspective i.e. the stereoisomer, which it represents is also symmetrical. The symmetry or dissymmetry of a molecule (stereoisomer) therefore, follows from the plane of symmetry test as applied to its Fischer projection diagram.

The use of projection diagrams for the determination of stereo-isomer number necessitates the comparison of the various diagrams which can be constructed. The comparison procedure consists simply in moving the diagrams, without rotation, in the plane of the paper and attempting to superpose them. If two projections are superposable then they are identical and represent the same stereoisomer, however, the non-superposability of projections does not mean necessarily that the projections are different. This problem is resolved by the application of some further properties of projection diagrams followed by a further test for superposability, if the diagrams are still non-superposable then they do represent different stereoisomers. The properties referred to relate to (*a*) the effect on the meaning of a projection when it is rotated and (*b*) to the effect on meaning when groups are interchanged.

The rotation of a Fischer diagram in the plane of the paper can change its meaning; thus a rotation of 90 degrees (for a monocentred system) converts the projection into its enantiomeric one; a rotation of 180 degrees in the plane (for mono- or polycentred systems) does not change the meaning of the projection. The special importance of 90 degree or 180 degree rotations will become clear in subsequent discussion. Rotations of Fischer type projections out of the plane of the paper are disallowed.

Group interchanges also can change the meaning of a projection diagram; an odd number of interchanges of groups about an asymmetric centre on the projection inverts ('enantiomerizes') that particular centre; if two or an even number of group interchanges are made then the projection remains unchanged in meaning and represents the same stereoisomer as before.

$$
\begin{array}{ccc}
\text{CHO} & \text{H} & \text{CHO} \\
\text{HO}\!-\!\!\!-\!\text{H} & \text{OHC}\!-\!\!\!-\!\text{CH}_2\text{OH} & \text{HO}\diagdown\diagup\text{H} \\
\text{CH}_2\text{OH} & \text{OH} & \text{CH}_2\text{OH} \\
\text{XXIV} & \text{XXV} & \text{XXVI}
\end{array}
$$

$$
\begin{array}{cc}
\text{H} & \text{CH}_2\text{OH} \\
\text{OHC}\diagup\diagdown\text{CH}_2\text{OH} & \text{H}\!-\!\!\!-\!\text{CHO} \\
\text{OH} & \text{OH} \\
\text{XXVII} & \text{XXVIII}
\end{array}
$$

The derivation of some of the above properties may be illustrated by glyceraldehyde, thus a 90 degree rotation of the projection (XXIV), in the plane of the paper, gives projection (XXV); the perspective diagrams corresponding to these projections are (XXVI) and (XXVII) respectively; a comparison of these perspectives shows them to be enantiomers. Two group interchanges (H and CHO then CHO and CH₂OH) on the projection diagram (XXV) gives the new diagram (XXVIII). These projections clearly are non-superposable and at first sight appear to represent different stereoisomers, however, a comparison of perspectives shows that, in fact, they are identical and represent the same stereoisomer. The other projection diagram properties quoted may similarly be verified.

The determination of stereoisomer number by the general permutation method may be discussed now for the case of glyceraldehyde. In the use of the method the outline projection diagram is drawn and then all possible permutations of groups are made on the diagram.

The projections (XXIX)–(XXXIII) are some of the possible permutations; these projections cannot be superposed no matter how they are moved in the plane of the paper and it might be concluded that at least five stereoisomeric glyceraldehydes exist. However, a reference to the properties of projections shows that this is not the case since some of the diagrams are identical. Thus, the projections (XXIX) and (XXX) are mirror images and are

dissymmetric hence they represent enantiomers; as indicated earlier the property of dissymmetry as well as that of non-superposable mirror images must be invoked if projections are to be enantiomeric. A rotation of (XXXI) through 90 degrees makes it superposable on (XXIX) and so the two projections are enantiomers i.e. (XXXI) is identical to (XXX). Two group interchanges on (XXXII) (H and OH then H and CH$_2$OH) gives (XXIX) and so these two projections represent the same stereoisomer. In the same way the projections (XXXIII) and (XXX) can be shown to be identical and, in fact, all other permutations which can be drawn can be shown to be identical to either (XXIX) or (XXX). One concludes that only two different projection diagrams can be drawn for glyceraldehyde and that these represent a pair of enantiomers. In practice, the general permutation procedure can be simplified to some extent by keeping in mind the number of isomers permitted by the 2^n rule; secondly, having made the first arbitrary permutation it should be tested for symmetry, if dissymmetric the enantiomeric projection should be drawn; this procedure should be repeated with each new projection. Thus, with glyceraldehyde only two enantiomers are possible (by the 2^n rule). The arbitrary permutation (XXIX) is drawn, it is found to be dissymmetric and so the enantiomer (XXX) is drawn, clearly no further permutations need to be considered.

K

The general permutation method for higher systems is more tedious than the group specification method but, as will be shown later, it has a stereochemical value not found with the group specification method. The application of the general permutation method to the special molecules such as glyceraldehyde gives projections such as (XXXI) for an enantiomer; this form of projection is frequently, but mistakenly described as a Fischer projection; it will be clear from earlier discussion that these are not Fischer projections proper but, of course, may be called Fischer type projections.

MOLECULES HAVING TWO ASYMMETRIC CENTRES

MOLECULES WITH SIMILAR CENTRES

The most fundamental problem associated with the stereoisomerism of a compound is the number of stereoisomers possible and their geometry. The discussion of monocentred systems used only the Fischer projection method because modern and Newman projections are illustrated best by higher systems; all these methods will be used in discussing bi- and polycentred systems.

Fischer Projections

Tartaric acid (XXXIV) is an historically important bicentred system having similar centres. The selection of a perspective and its arrangement is a more complex problem for a bicentred system than it is for a monocentred system and the procedure involves two separate steps.

COOH
|
CHOH
|
CHOH
|
COOH

XXXIV

HOOC H 〉 COOH OH / H 〉 OH

XXXV

HO 〉 H HOOC 〉 COOH / H 〉 OH

XXXVI

The tartaric acid molecule has an ethane carbon skeleton and on classical theory it may have a specific geometry or internal rotation involving several conformations may occur. This sort of problem does not occur, of course, with monocentred systems since these have the fixed tetrahedral distribution of valencies. The second problem is that having selected a particular geometry as the perspective this has to be arranged on the paper in order to derive a projection diagram.

Experimentally, it was known that one, and only one, optically inactive tartaric acid exists. This isomer may exist in a specific conformation but if it does then its geometry must be symmetrical since the isomer is optically inactive. The conformations (XXXV) and (XXXVI) are the only symmetrical ones possible and further, they are symmetrical for quite different reasons, thus (XXXV) has a plane of symmetry whereas (XXXVI) has a centre of symmetry. The existence of two possible symmetrical perspectives introduces the first selection problem since each will give a different projection diagram for the same stereoisomer. An arbitrary selection of either (XXXV) or (XXXVI) is necessary therefore and Fischer chose the eclipsed form (XXXV).

The above described classical selection procedure assumes that the optically inactive stereoisomer of tartaric acid adopts a specific conformation. However, if classical free rotation occurs the molecule might exist in both conformations (XXXV) and (XXXVI) as well as others. The question arises, therefore, as to whether a projection diagram derived from a quite arbitrary perspective which may not even exist in reality, has any meaning. In the

classical period a direct answer to this question was impossible; Fischer recognized that projection diagrams for higher systems were artificial devices whose justification was that they sufficed as a means of representing stereoisomers. It is now known that Fischer perspectives and projections for higher systems are, in fact, in error and an understanding of Fischer's success has only become apparent with the development of modern theory.

The perspective (XXXV) selected by Fischer may be arranged in the paper, for projection, in a variety of ways. The drawings (XXXVIa)–(XXXIX) represent some of the possible arrangements of (XXXV); these drawings are to be interpreted as meaning that the ethane carbon atoms are arranged in the plane of the paper with the other groups projecting above or below this plane. In the actual diagrams the carbon-carbon bond is drawn, for ease of viewing, at an angle to the edge of the page but for deriving the projection diagram the perspective arrangement is to be imagined so that the carbon-carbon bond is parallel to the edge of the page. Fischer arbitrarily selected the perspective arrangement (XXXVIII) and so derived the projection diagram (XL) to represent the optically inactive stereoisomer of tartaric acid.

A variety of perspectives and arrangements is similarly possible for the two enantiomeric tartaric acids which make up the total of possible stereoisomers for this substance. However, it is obviously best to select and arrange perspectives for these isomers so that their projections are as similar as possible to the one derived for the inactive isomer. Fischer used the arrangements (XLI) and (XLII) and the diagrams (XLIII) and (XLIV) are, therefore, the Fischer projections for the enantiomeric tartaric acids.

The use of Fischer projections for the determination of stereo-isomer number for a bicentred system follows that previously described for monocentred systems. The group specification method requires the selection of two groups which are attached to the vertical of the projection diagram outline and the remaining groups are permutated then, about the horizontals. In the case of tartaric acid the carboxyl groups are chosen for attachment to the vertical since this was the choice Fischer made. The permutation of hydrogen and hydroxyl groups about the horizontals then gives the four different projection diagrams (XLV)–(XLVIII). These four

XLV	XLVI	XLVII	XLVIII

different permutations may or may not represent different stereo-isomers whether or not they do must be determined now by the application of projection diagram properties. A rotation of projection (XLVI) through 180 degrees, in the plane of the paper, makes it superposable on the diagram (XLV) and so the two are identical. The projections (XLVII) and (XLVIII) cannot be made superposable and also are different to (XLV) and represent, therefore, two further stereoisomers. The isomer (XLV) is symmetrical and inactive whereas the isomers (XLVII) and (XLVIII) are dissymmetric, optically active and enantiomeric. Thus, a complete analysis of the stereoisomers of an aliphatic compound using the group specification method requires the use of the symmetry property, the enantiomer property and the 180 degree rotation property of projection diagrams.

The general permutation method requires the use of all the projection diagram properties. The diagrams (XLIX)–(LIV) are some of the possible permutations for the case of tartaric acid and at first sight they appear to be different and so represent different stereoisomers. However, a rotation through 180 degrees of (L) gives (XLIX); the diagram (LI) is not superposable on either (L) or (XLIX) nor can it be made so without changing its meaning, obviously it represents another isomer. The diagram (LII) is the dissymmetric non-superposable mirror image of (LI) and so is the enantiomer of (LI); the diagram (LIII) may be made identical to (L) by two group interchanges (OH and COOH attached to the lower carbon centre, then OH and H). In a similar way the group interchange process shows that (LIV) is identical to (LII); any other projection diagrams can be shown to be identical to (XLIX), (LI) or (LII) and so these represent the only three stereoisomers

```
     COOH              COOH              COOH
      |                 |                 |
 H————+————OH     HO————+————H       H————+————OH
      |                 |                 |
 H————+————OH     HO————+————H      HO————+————H
      |                 |                 |
     COOH              COOH              COOH

     XLIX                L                 LI


     COOH              COOH              COOH
      |                 |                 |
HO————+————H      HO————+————H      HO————+————H
      |                 |                 |
 H————+————OH      H————+————COOH   HO————+————COOH
      |                 |                 |
     COOH               OH                H

      LII               LIII              LIV
```

of tartaric acid. As pointed out previously, in using the general permutation method it is useful to bear in mind the 2^n rule since this can reduce the tedium of drawing out diagrams.

The optically inactive stereoisomer of tartaric acid is rather special since although it is inactive it does contain asymmetric centres. This fact again emphasizes that molecular symmetry properties and not asymmetric centres are the fundamental factors associated with optical stereoisomerism. Molecules containing asymmetric centres but having overall symmetry and optical inactivity are called meso forms. However, it is to be realized that although a substance may have symmetrical stereoisomers these are not meso forms unless they contain asymmetric centres; thus the trihydroxyglutaric acid (XXX) (Chap. 5) is a meso form but the isomer (XXXV) (Chap. 5) is not; this latter isomer is in fact, not an optical stereoisomer at all. It will be clear that meso forms must contain even numbers of asymmetric centres and that these are distributed as enantiomeric pairs about some plane of symmetry within the molecule; alternatively the molecule may have a centre of symmetry or a fourfold alternating axis.

In a group of optical stereoisomers any two, which are not

enantiomers, are called diastereoisomers. Enantiomers are identical (for present purposes) in all properties except rotatory power whereas diastereoisomers have quite different properties. Thus, the enantiomeric tartaric acids have a m.p. of 170° and $[\alpha]_D^{20} \pm$ 11·98° (in water) but the diastereoisomeric meso form melts at 206° and is optically inactive.

Modern Projections
Fischer projections for bicentred systems derive from the geometrically unreal eclipsed ethane conformation. Conformational theory and x-ray evidence show that molecules of meso-tartaric acid in the crystal have the staggered ethane conformation; this was the alternative rejected by Fischer in his arbitrary choice of starting perspective. It is of interest that the staggered conformation for the tartaric acids was suggested, on chemical grounds, as early as 1931 (Amadori[148]) but its significance in conformational terms has been recognized only recently. The geometrically correct perspective diagram, therefore, for meso-tartaric acid is (LV) and (LVI)– (LVII) are the perspectives of the enantiomers. A knowledge of the

actual geometry of molecules clearly obviates the need for the arbitrary selection of a perspective but, of course, in the derivation of a projection the problem of perspective arrangement still exists. Modern projections are derived from modern perspectives by arranging the latter in the paper in a manner similar to that described for Fischer perspectives; the diagrams (LV)–(LVII) show these arrangements. The projection of these arrangements on to the plane of the paper gives the modern diagrams (LVIII)– (LX) for the stereoisomeric tartaric acids, clearly, they represent different stereoisomers than those of the corresponding Fischer projections. A further difference in modern and Fischer projections is, of course, their geometrical meaning, thus, in the modern bicentred projection the lower part of the vertical, for example, refers to a part of the perspective which projects above the plane of the paper while in a Fischer diagram this part of the corresponding perspective projects below the plane. These differences in geometrical meaning must be kept in mind when using these types of projections.

Modern projections may be used to determine stereoisomer number for a substance in either of the two ways described previously. However, the use of modern projections requires a different set of

properties to those applicable to Fischer type projections; these properties of modern diagrams also depend on whether the compound has an odd or an even number of asymmetric centres. Thus (*a*) the symmetry properties of projections of odd numbered systems are determined by the plane of symmetry test whereas the properties of even systems are detected by the centre of symmetry test; (*b*) for odd systems a rotation of the projection through 180 degrees in the plane of the paper leaves the diagram unchanged but with even systems a rotation of 180 degrees out of the plane and about the broken line as axis illustrated in (LVIII) must be considered, such a rotation leaves the projection unchanged; (*c*) for any system an odd number of group interchanges about a centre inverts that centre while an even number of interchanges leaves the projection unchanged.

Fischer projection diagrams, for historical reasons, are the ones in general use in spite of their having incorrect geometrical meaning. Modern projections are as easy to draw and remember as are the Fischer type and have the very great advantage of deriving from geometrically real perspectives, this fact results in a correct representation by the projection of the symmetry properties of the particular stereoisomer and a correct general indication of the actual molecular geometry. Thus, the modern projection diagram for meso-tartaric acid has a centre of symmetry and shows the hydroxyl groups as distant from each other; just these properties are found in the actual molecule of meso-tartaric acid.

It was pointed out earlier that classical theory was unable to analyse the implications of the concept of free rotation for projection diagrams. This matter is of the greatest importance since a Fischer projection for an optically inactive isomer would be shown as a symmetrical drawing whereas, in fact, such an isomer may be made up of a pair of rotationally interconvertible enantiomers. The staggered form of meso-tartaric acid, from which the modern projection is derived, is present only in the crystal; in the melt or in solution other rotational isomers exist and it becomes relevant

LXI   LXII   LXIII

to inquire as to their implications for modern projections. Conformational theory shows that in an appropriate environment, such as solution, meso-tartaric acid exists as an equilibrium mixture of five optical rotational isomers. The isomer (LXI) is the symmetrical one discussed previously and the isomers (LXII)–(LXIII) are dissymmetric rotational derivatives of (LXI); these dissymmetric forms exist, of course, as enantiomers so making a total of five meso forms. The sizes of the substituents are in the order COOH > OH > H and, therefore, in terms solely of the steric factor the symmetrical isomer (LXI) is the most stable one. The steric interactions in the other rotational forms show them to have identical free energies although, of course, these are higher than the energy of (LXI). These relative values of free energy for the meso isomers account for the crystalline meso acid being made up entirely of the conformations (LXI).

In solution, the form (LXI) predominates and the other four conformations are present in equal amounts; thus, such a solution is optically inactive in spite of the presence of dissymmetric conformations. The optical inactivity of crystalline meso-tartaric acid is, of course, due to the substance being made up purely of symmetrical molecules. The modern projections of the secondary meso isomers (LXI)–(LXIII) are shown as (LXIV)–(LXVI) respectively. Thus, modern theory advances classical theory considerably since it is able to decide on the stable conformations a molecule can have and so all the meaningful projection diagrams may be derived. The projection diagrams such as (LXIV)–(LXVI) represent a group of secondary isomers, however, it is frequently necessary to discuss only the primary form equivalent to the group of secondary isomers and it is best to do this by selecting the most stable conformation. Thus the diagram (LXIV) is selected to represent the primary meso-tartaric acid.

$$
\begin{array}{ccc}
\text{COOH} & \text{COOH} & \text{COOH} \\
\text{HO}{-}\!|\!{-}\text{H} & \text{HO}{-}\!|\!{-}\text{H} & \text{HO}{-}\!|\!{-}\text{H} \\
\text{H}{-}\!|\!{-}\text{OH} & \text{HO}{-}\!|\!{-}\text{COOH} & \text{HOOC}{-}\!|\!{-}\text{H} \\
\text{COOH} & \text{H} & \text{OH} \\
\textbf{LXIV} & \textbf{LXV} & \textbf{LXVI}
\end{array}
$$

The projection diagrams for secondary stereoisomers are of interest in connection with the general permutation method. Thus, it will be apparent that in the determination of the stereoisomer number, for tartaric acid, by general permutation, the diagrams (LXIV)–(LXVI), among others, arise naturally. It will be recalled that in this method all projection permutations are drawn and then examined for equivalence by applying the properties of projections. This procedure shows that the drawn projections are made up of groups of equivalent projections and each equivalent group represents one primary isomer. The significance of these groups of identical projections is now apparent from conformational theory; thus each group represents rotational isomers although the actual number of different rotational forms is always less than the number of identical projections in the group, since some of the permutations represent the same rotational isomer.

One concludes from the above discussion, that the best method for the determination of stereoisomer number for a bicentred system is by a combination of the group specification and general permutation methods. The group specification method quickly gives the primary stereoisomers and each of these is then subjected to a limited form of general permutation by an exchange of groups about one (e.g. the lower) asymmetric centre. The group exchange is made in one particular direction, for example anti-clockwise, and this corresponds to two group interchanges of the type mentioned earlier and so the projection remains unchanged; thus (LXV) and (LXVI) derive from (LXIV) by an anti-clockwise group exchange. The final step consists in examining the symmetry of the projections deriving from a symmetrical projection and drawing the mirror images of the dissymmetric ones, thus, group specification gives

(LXIV) for meso-tartaric acid; group exchange gives (LXV) and (LXVI), these latter are dissymmetric and drawing their mirror image projections completes the analysis. When the starting projection is dissymmetric the group exchange suffices to determine the stereoisomers; thus (LXVII) is an enantiomer; group exchange

$$
\begin{array}{ccc}
\text{COOH} & \text{COOH} & \text{COOH} \\
\text{H}\!-\!\!-\!\text{OH} & \text{H}\!-\!\!-\!\text{OH} & \text{H}\!-\!\!-\!\text{OH} \\
\text{H}\!-\!\!-\!\text{OH} & \text{HO}\!-\!\!-\!\text{COOH} & \text{HOOC}\!-\!\!-\!\text{H} \\
\text{COOH} & \text{H} & \text{OH} \\
\text{LXVII} & \text{LXVIII} & \text{LXIX}
\end{array}
$$

gives (LXVIII)–(LXIX) and these are the only rotational forms possible for this enantiomer. The fact that modern projections provide a means of representing secondary stereoisomers is clearly a further advantage of this type of projection over the classical Fischer type.

Newman Projections

This type of projection provides a third method of representing stereoisomers. Newman diagrams are specifically useful for bicentred systems or for discussing any two adjacent centres in other systems including alicyclic molecules. The starting perspective for bicentred systems is, of course, the staggered ethane carbon skeleton. This perspective is arranged so that the plane of the paper passes through the mid-point of the carbon-carbon bond and cuts this bond normally. The projection of the ethane carbon skeleton on to the plane of the paper then gives the Newman projection diagram outline (LXX). It will be clear that Newman diagrams correspond to a view of the molecule as seen by looking along the ethane carbon-carbon bond; the full lines of the diagram represent the valencies of the carbon atom nearest to the eye. The use of full and broken lines in the diagrams must not be considered to have geometrical meaning in the projection; the projection is a planar drawing and has no geometrical significance beyond being planar; the full and broken lines are introduced merely for convenience in using the diagrams. Fischer and modern projections correspond to that view obtained by looking down on the molecule; these latter two types of projection show all the important bonds in the molecule while in Newman diagrams the ethane carbon-carbon bond is not seen.

Newman diagrams may be used to determine the stereoisomer number for a substance by the combined method described for modern projections. The group specification procedure gives the primary isomers and therefore in Newman, as in modern projections, the largest groups are the ones chosen for attachment to the verticals.

LXX   LXXI   LXXII

LXXIII   LXXIV

This choice results in the primary isomers being represented by the most stable conformation. The COOH groups, therefore, are chosen in order to represent the tartaric acids and fortuitously, the resulting projections are analogous to Fischer's classical projections. The permutation of the remaining groups gives the diagrams (LXXI)–(LXXIV). A comparison of these projections for equivalence may be made by the use of a similar set of rules to those described for modern projections, however, it will be clear from previous discussion that for the particular purpose of determining stereoisomer number certain simplifications are possible. Thus, the projections (LXXIII) and (LXXIV) are mirror images and also are symmetrical (centre of symmetry) hence they are equivalent; this conclusion could have been drawn also by using the 180 degree rotation rule. The permutations (LXXI) and (LXXII) are mirror images but are dissymmetric and so represent enantiomers.

The primary stereoisomers represented by (LXXI)–(LXXIII) may be converted now into their groups of secondary forms as described earlier; (LXXI) gives (LXXV)–(LXXVI) and (LXXII) gives (LXXVII)–(LXXVIII); the meso isomer (LXXIII) gives (LXXIX)–(LXXX) and their enantiomers (LXXXI)–(LXXXII). The meso isomer (LXXIII) is the meso-transoid form while the other meso forms have their COOH groups in a cisoid relationship.

The above discussion has dealt with the determination and the representation of stereoisomers by means of projection diagrams.

The projection diagram has definite value for the reasons stated earlier but it is to be appreciated that it is not the only means of determining and representing the stereoisomers of a substance. Thus, as has been implied previously, the perspective diagram provides an alternative method and in fact is the method generally

COOH
HO.·|.·COOH
H⟍⟋OH
H

LXXV

COOH
HOOC.·|.·H
H⟍⟋OH
OH

LXXVI

COOH
H.·|.·COOH
HO⟍⟋H
OH

LXXVII

COOH
HOOC.·|.·OH
HO⟍⟋H
H

LXXVIII

COOH
H.·|.·COOH
H⟍⟋OH
OH

LXXIX

COOH
HOOC.·|.·OH
H⟍⟋OH
H

LXXX

COOH
HOOC.·|.·H
HO⟍⟋H
OH

LXXXI

COOH
HO.·|.·COOH
HO⟍⟋H
H

LXXXII

used for alicyclic systems. The perspective method is clearly independent of molecular complexity whereas for these complex systems the construction of useful projections is difficult or impossible. The perspective method requires knowledge of the actual or hypothesized geometry of a molecule's carbon skeleton; the method then permutates substituents on this skeleton and determines which of the perspectives, so derived, are equivalent by considering possible internal rotations and vibrations as well as the simple procedure of attempting superposition. The stereoisomers of a compound derive naturally from the perspective permutation method and further, the method requires no knowledge of symmetry properties or of the presence of asymmetric centres; symmetry properties need be applied only subsequently in order to define the optical properties of the stereoisomers and to pair off enantiomers. Although the concepts of asymmetric centres and symmetry properties are not necessary to the determination of stereoisomer number they are, of course, still useful and provide a simple means of detecting optical stereoisomerism in molecules. Perspective permutation will be exemplified later for alicyclic molecules.

MOLECULES WITH DIFFERENT CENTRES

The sugar (LXXXIII) is an example; it exists as $2^2 = 4$ primary stereoisomers made up of $2^2/2$ pairs of enantiomers. These isomers may be derived by any of the three projection methods described, or alternatively by the perspective method. The Fischer projections are obtained by attaching the CHO and CH_2OH groups to the verticals of the projection diagram outline and permutating the remaining groups. This application of the group specification

| | LXXXIII | LXXXIV | LXXXV | LXXXVI | LXXXVII |

method gives the four projections (LXXXIV)–(LXXXVII); they are quite different and represent the possible primary stereoisomers of (LXXXIII).

The combined method, using modern projections, gives the twelve diagrams (LXXXVIII)–(XCIX) and these show that the substance exists as four primary isomers with each providing a

group of three secondary rotational forms. The projections (LXXXVIII), (XCI), (XCIV) and (XCVII) represent the most stable conformations of each group and by the convention indicated earlier these are used to represent the primary isomers when interest is limited to these.

The sugars represented by the Fischer projections (LXXXIV) and (LXXXV) are called erythroses; those represented by (LXXXVI) and (LXXXVII) are called threoses. These names have been introduced into stereochemical nomenclature as a general means of describing bicentred stereoisomers having a substitution

$$
\begin{array}{ccc}
\text{CHO} & \text{CHO} & \text{CHO} \\
\text{HO}\!-\!\!-\!\text{H} & \text{HO}\!-\!\!-\!\text{H} & \text{HO}\!-\!\!-\!\text{H} \\
\text{H}\!-\!\!-\!\text{OH} & \text{HO}\!-\!\!-\!\text{CH}_2\text{OH} & \text{HOH}_2\text{C}\!-\!\!-\!\text{H} \\
\text{CH}_2\text{OH} & \text{H} & \text{OH} \\
\text{XCIV} & \text{XCV} & \text{XCVI}
\end{array}
$$

$$
\begin{array}{ccc}
\text{CHO} & \text{CHO} & \text{CHO} \\
\text{H}\!-\!\!-\!\text{OH} & \text{H}\!-\!\!-\!\text{OH} & \text{H}\!-\!\!-\!\text{OH} \\
\text{HO}\!-\!\!-\!\text{H} & \text{H}\!-\!\!-\!\text{CH}_2\text{OH} & \text{HOH}_2\text{C}\!-\!\!-\!\text{OH} \\
\text{CH}_2\text{OH} & \text{OH} & \text{H} \\
\text{XCVII} & \text{XCVIII} & \text{XCIX}
\end{array}
$$

similar to these sugar molecules. Thus, the Fischer type projections for two stereoisomers of the substance $CH_3CHBrCHBrC_6H_5$ are (C) and (CI); the projection (C) has two sets of identical substituents and so represents an erythro isomer; the projection (CI) represents a threo form. The modern projections for erythro and

$$
\begin{array}{cc}
\text{C}_6\text{H}_5 & \text{C}_6\text{H}_5 \\
\text{H}\!-\!\!-\!\text{Br} & \text{H}\!-\!\!-\!\text{Br} \\
\text{H}\!-\!\!-\!\text{Br} & \text{Br}\!-\!\!-\!\text{H} \\
\text{CH}_3 & \text{CH}_3 \\
\text{C} & \text{CI}
\end{array}
$$

threo isomers are, of course, different to the Fischer projections; thus (XCIV) and (XCVII) are erythroses and (LXXXVIII) and (XCI) are threoses.

The briefly mentioned problem of the stability of the rotational isomers of a bicentred system may be considered now in its general form. The substituents of these systems may be described, in steric terms, as large (L), medium (M) and small (S). The compound (CII) has four stereoisomers one of which is presented by the Newman projection (CIII); the projections (CIV) and (CV) are the rotational relatives of (CIII). These Newman diagrams clearly show that, in terms of the steric factor, the stability order is

| CII | CIII | CIV | CV |

(CV) < (CIV) < (CIII). This analysis permits the deduction of the relative stabilities of the secondary isomers of a group and is the basis of the convention for representing primary isomers in modern or Newman projection form.

Other Molecules having Asymmetric Centres

The compound trihydroxyglutaric acid is an example; the earlier analysis of the stereoisomerism of this substance was based on an examination for àsymmetric centres. The analysis showed trihydroxyglutaric acid to be a rather special compound since, of the four primary stereoisomers, two isomers have three asymmetric centres present and the other two have two such centres. The determination of the stereoisomer number of substances such as trihydroxyglutaric acid is considerably simplified by the application of the projection diagram method.

| CVI | CVII | CVIII | CIX |

The group specification method readily gives the four different Fischer projections (CVI)–(CIX) and therefore, these represent the primary stereoisomers of trihydroxyglutaric acid. The symmetrical isomers (CVI) and (CVII) are the optically inactive meso

CX

CXI

CXII

CXIII

CIV

CV

forms and (CVIII)–(CIX) is the pair of enantiomers. The Fischer
perspectives from which the above projections derive may be
drawn as (CX)–(CXIII); the modern perspectives for the stereoisomers may be drawn as (CIV)–(CVII). Modern perspectives
for higher aliphatic systems are best drawn with a zig-zag aliphatic
chain; since this generally represents the most stable skeletal
conformation. A polycentred perspective is easiest to view as a

drawing with the aliphatic carbon chain in the plane of the paper and with other valencies projecting behind (broken line) or in front (full line) of this plane. In order to derive the modern projections from these perspectives they are arranged with the carbon

CVI   CVII

chain parallel to the edge of the page and with C_1 nearest to the top of the page; the chain is placed so that the carbon atoms lie in a plane perpendicular to that of the paper and so that the plane of the paper passes through the mid-points of the carbon-carbon bonds. This arrangement results in the end group (e.g. COOH) attached to C_1 always being below the plane of the paper; the other end group is below the plane when the aliphatic chain has an odd number of carbons and above the plane when this number of atoms is even.

CVIII   CIX

The projection of the arranged modern perspectives of trihydroxyglutaric acid isomers gives diagrams similar to Fischer projections but, as was shown for the tartaric acids, these similar projections refer to quite different stereoisomers. The above discussion shows that in the determination of stereoisomer number the existence of asymmetric centres can be ignored; the substance (CVIII) is a further example. This compound has the complication of a non-asymmetric carbon atom intercalated between asymmetric centres; the procedure is as usual, the projection outline is drawn and the substituents permutated; the diagram (CIX) represents one such permutation.

The aliphatic form of glucose is an example of a polycentred system; the Fischer perspective and projection are (CX) and (CXI) respectively; the modern equivalents of these are (CXII) and

CX

CXI

CXII

CXIII

(CXIII). The determination of stereoisomer number follows the usual procedures; the combined method, as before, gives both primary and secondary isomers. The examples discussed above all contain asymmetric centres as part of a straight aliphatic chain; compounds having asymmetric centres in branching chains or having heteroatoms present may also be analysed by the projection diagram methods.

7

OPTICAL STEREOISOMERISM:
(III) ALICYCLIC MOLECULES HAVING
ASYMMETRIC CENTRES

INTRODUCTION

The occurrence of optical stereoisomerism in the alicyclic series is analogous to that in the aliphatic series; alicyclic primary stereoisomers may exist due to the presence of asymmetric centres or because of general molecular dissymmetry. Monocyclic compounds containing an even number of ring carbon atoms can also show *cis-trans* isomerism in addition to optical stereoisomerism. The present discussion concerns only alicyclics having asymmetric centres. Optical stereoisomerism in alicyclic compounds may be primary and secondary; the classical theory dealt successfully with the former type but secondary isomerism requires modern conformational analysis.

In the classical period the stereoisomerism of monocyclic systems, with the exception of sugars, was interpreted by means of Baeyer planar models or their perspective diagrams. Polycyclic molecules were not extensively considered in the early classical period because the Baeyer models for even the simple polycyclic compound, decahydronaphthalene, show superstrained isomers whose existence was doubtful. A start on these molecules was made by Mohr[133] but a detailed analysis has become possible only with the development of conformational theory.

Sugar molecules are now known to be stereochemical analogues of cyclohexane and cyclopentane but in the classical period this was not presupposed and these molecules were represented by the original Fischer perspectives and projections. In 1926, these two methods of representing cyclic systems were correlated by Drew and Haworth[149] who showed how Fischer perspectives and projections may be converted to Baeyer-type perspectives. This theoretical correlation implied a stereochemical relation between sugar molecules and alicyclic molecules proper; this has been shown since to be fully justified.

It is now well established that the planar Baeyer perspectives are valid only for cyclopropane and its derivatives; other alicyclic molecules have non-planar molecules. In spite of this it is clear that Baeyer perspectives are much closer to geometrical reality than are Fischer perspectives. However, in the classical period the Baeyer and Fischer methods sufficed because they were able to predict and interpret the primary stereoisomer number for alicyclic substances. The development of conformational theory has not altered the classical conclusions as to primary stereoisomer number for a substance but correctly analyses the symmetry properties of the stereoisomers and includes secondary isomers.

As mentioned earlier the stereoisomer number for an alicyclic substance is determined by the perspective permutation method; using a Baeyer perspective only the primary isomers can be determined but with a modern perspective all the isomers are obtained. Projection diagrams for alicyclic systems can be constructed but except in special cases offer no advantages for the determination of isomer number or the representation of stereoisomers. The various types of alicyclic system may now be discussed.

MONOCYCLIC MOLECULES

A monosubstituted cyclopropane is unique and is incapable of primary stereoisomerism. However, appropriate further substitution can give cyclopropanes having one, two or three ring carbons as asymmetric centres. The origin of asymmetric centres in alicyclic compounds follows from the symmetry analysis presented for aliphatic molecules. Thus, Baeyer cyclic molecules are constructed from distorted tetrahedral centres and the nature of the distortion happens to be such that appropriately substituted centres have the same capacity to produce asymmetry as have normal tetrahedral centres.

The bromine substituted carbon atom in the cyclopropane derivative (I) is an asymmetric centre. The detection of such centres by the inspection of the Kekulé-Couper structure is a slight modification of the 'four different substituents' test applied to aliphatics. Thus, considering the structure (I) one starts at the bromine substituted centre and proceeds round the ring; in one direction a CH_2 and then a CCl_2 group are encountered while in the other direction the same groups are found but in a different order, hence the starting centre is asymmetric as the two different paths correspond to two different substituents. The asymmetric centre results in (I) existing as a pair of enantiomers; their Baeyer perspectives are (II) and (III). The Baeyer perspectives are drawings of the Baeyer model arranged with its ring inclined at an angle to the plane of the paper of somewhat less than 90 degrees. This form of perspective makes the three-dimensional significance of the drawings easily seen particularly when the frontal ring bond is drawn as a heavy line. These perspectives may be simplified, as in (IV) and (V), by drawing the exacyclic valencies perpendicular to

the ring. This simplification facilitates drawing and, more important, makes clear the stereochemically important property of substituents as lying above or below the ring plane. This 'vertical' perspective is fundamentally justified by the fact that the symmetry properties of the 'tetrahedral' units are still preserved in spite of the increased distortion.

As pointed out the stereoisomer number for an alicyclic system is determined usually by the perspective permutation method. Thus, for cyclopropane dicarboxylic acid, the parent perspective is drawn and then all the different permutations of the substituents made; some of the permutations possible for this compound are (VI)–(IX). The final step in the procedure is the testing of the permutations for equivalence by simple superposition and by considering the possibilities of internal rotation. The perspectives (VI) and (VII) can be superposed and are equivalent, the other two represent enantiomers; the test of internal rotation actually does not apply to cyclopropane derivatives since a three-membered ring is a rigid structure. Other perspective permutations are possible, of course, but since the Baeyer cyclopropane ring is an equilateral triangle any other permutations are identical to (VII), (VIII) or (IX); a non-equilateral structure would increase the number of stereoisomers.

The number and type of primary stereoisomers of an alicyclic substance corresponds to that for the analogous aliphatic problem. The cyclopropane dicarboxylic acid has two similar asymmetric centres and exists as the meso (cisoid) form (VII) and as the pair of enantiomers (transoid) (VIII)–(IX). However, in the analogous tartaric acid problem secondary isomers exist whereas for the cyclopropane dicarboxylic acid this form of isomerism is not shown. Compounds in which a heteroatom replaces a methylene group of the cyclopropane ring may be treated as described above; a substituted ethylene oxide can have only two asymmetric centres but a substituted ethyleneimine (X) can have three such centres. The asymmetric nitrogen atom in ethyleneimine derivatives is associated historically with the problem of the primary isomerism of substituted p^3 nitrogen. Thus, it was considered that hindrance to inversion would exist in ethyleneimines and an activation energy of 38 kcal/mole has been calculated (Kincaid and Henriques[130]). The inversion in these molecules is expected to be more difficult because the nitrogen atom uses p^3 orbitals in the imine and inversion requires a change to sp^2 hybrids with opening of the $\widehat{CNC}$; this process would destabilize further an already unstable system. Ethyleneimine derivatives have not yet been obtained as primary isomers based on asymmetric tricovalent nitrogen but it has been calculated that such isomers should be stable below $-50°$ (Bottini and Roberts[150]).

In the cyclobutane series the 1,2-disubstituted derivatives have asymmetric ring carbons but the 1,3 derivatives have none; this latter type of derivative, however, shows *cis-trans* isomerism. Cyclobutanes, appropriately substituted, can have up to four asymmetric centres. The above discussion is based on the classical perspectives for three and four membered rings but since these are closely similar to the actual molecular geometries the classical analysis is essentially identical to the modern analysis.

The cyclopentane molecule has a non-planar ring and this results in more stereoisomers, of different symmetry properties, than predicted by classical theory. Thus, cyclopentyl bromide (XI) is described classically by the Baeyer perspective (XII) and clearly, since the ring is a regular pentagon, only one perspective for the

XI   XII   XIII   XIV   XV

XVI   XVII   XVIII   XIX

XX   XXI   XXII

molecule can be drawn. However, the permutation of the bromine substituent on the modern puckered perspective gives the different diagrams (XIII)–(XXII); these ten non-superposable diagrams may or may not represent primary stereoisomers and this is now tested by considering possible internal rotations. Internal rotation in the classical planar models of cyclic compounds is not possible, of course, but rotation in puckered cyclopentane does occur and allows each ring carbon, in turn, to leave the ring plane. This mobile puckering results in easy interconversion of the stereoisomers represented by the perspectives (XIII)–(XXII) and cyclopentyl bromide exists, therefore, as an equilibrium mixture of these secondary isomers. An examination of the symmetry properties of these isomers shows that they exist as four pairs of enantiomers together with the two symmetrical optically inactive conformations (XXI) and (XXII). The enantiomeric conformations owe their dissymmetry to the fact that the bromine substituted centres are quasi-asymmetric ones. The symmetrical conformations, however, are not meso forms since no asymmetric atoms are present. The conformation (XXII) is expected to be the most stable and to predominate in the equilibrium mixture; the enantiomers will be present in equal amounts since they have equal energies; thus the equilibrium mixture is optically inactive as found experimentally. One concludes that modern and classical theory concur as to primary stereoisomer number but that a fundamental analysis is possible only on modern theory.

XXIII   XXIV   XXV

XXVI   XXVII   XXVIII   XXIX   XXX

XXXI   XXXII   XXXIII   XXXIV   XXXV

XXXVI   XXXVII   XXXVIII   XXXIX   XL

XLI   XLII   XLIII   XLIV   XLV

The cyclopentane derivative (XXIII) has one asymmetric centre and the 2^n rule predicts a pair of enantiomers; their classical perspectives are shown as (XXIV) and (XXV). A modern analysis for isomers shows a much more complex situation; thus, the large number of different (non-superposable) permutations (XXVI)–(XLV) can be drawn. A consideration of mobile puckering shows that the stereoisomers fall into the two groups (XXVI)–(XXXV) and (XXXVI)–(XLV); the interconversion of isomers within each group occurs easily but no interconversion between the groups is possible, further, one group of perspectives is enantiomeric with the other and so the substance exists as ten pairs of enantiomers. The classical analysis shows only a pair of primary enantiomers; the modern analysis shows that each of these exists as an equilibrium mixture of ten secondary forms.

The primary stereoisomers of 1,2-dimethylcyclopentane are shown as the Baeyer perspectives (XLVI)–(XLVIII); these show that the substance exists as a meso form (XLVI) and as a pair of enantiomers (XLVII)–(XLVIII); this result is expected in view of

XLVI   XLVII   XLVIII

the presence of two similar asymmetric centres. The permutation of substituents on a puckered cyclopentane perspective increases the number of possible stereoisomers. A consideration of internal rotation shows that these stereoisomers fall into three groups; interconversion between the groups is impossible but within the groups the isomers are interconverted easily. Thus, modern theory shows three primary isomers, in agreement with classical theory, but in addition it reveals the existence of secondary isomers.

A further stereochemical problem arises with di- and poly-substituted alicyclic molecules namely the stereochemical correlation of the modern and classical perspectives for stereoisomers. Thus, the classical perspective for, say, the meso isomer (XLVI) is geometrically different to the modern perspective for this isomer and in particular the CH_3 groups in (XLVI) are cisoid whereas this relationship of substituents may not exist in the corresponding modern perspective. An examination of the symmetry properties of the groups of secondary isomers of 1,2-dimethylcyclopentane shows that the isomers of one group are enantiomeric with those of a second group and clearly these two groups must represent the enantiomers of the substance. The third group of isomers is made up of dissymmetric forms together with symmetrical forms and further, the dissymmetric conformations can be paired off as enantiomers. The equilibrium mixture of rotational isomers is

optically inactive and clearly corresponds to the classical meso
form. The perspectives (XLIX)–(LI) represent some of the conformational equivalents of the classical meso form and (LII) is one
rotational form of an enantiomer. The stereochemical correlation
of the classical and the modern perspectives is now possible and

| XLIX | L | LI | LII |

| LIII | LIV | LV | LVI |

is done by reference to the projected valency angles in the two types
of perspectives. The projected valency angle between CH_3 groups
in (XLIX) is shown in the partial projection (LIII) and is 0 degrees;
the substituents therefore are oriented truly cisoid as in eclipsed
2,3-dibromobutane; thus, at least, in this particular perspective the
CH_3 groups have the same geometrical relationship that they have
in the classical perspective. A projected angle of 0 degrees is shown
for (XLIX) but to be strictly accurate this angle is actually only
close to 0 degrees due to ring puckering, nevertheless the argument
is essentially valid. The conformations (L) and (LI) also have
projected angles of near 0 degrees shown in (LIV) and (LV) although
they are slightly greater than that in (XLIX); again an essentially
cisoid relationship of substituents exists and, in fact, is found in all
other conformations representing the inactive stereoisomer. The
angle in the isomer (LII) is 120 degrees shown in (LVI) and the
substituents may be defined as transoid; the Baeyer perspective
proper, for this enantiomer, also has the 120 degree angle but the
'vertical' Baeyer perspective has its substituents at the truly transoid
angle of 180 degrees. One concludes that for cyclopentane derivatives the orientation of substituents is essentially the same on both
classical and modern perspectives. A comparison of the relationships
of groups in alicyclic and aliphatic systems is of interest; thus in
2,3-dibromobutane cisoid substituents have an angle of 60 degrees;
the truly cisoid angle of 0 degrees is found only in the unreal eclipsed

161

conformation; transoid substituents have an angle of 180 degrees and clearly the cisoid-transoid relations in aliphatic systems are quite different to those found in cyclopentanes, however, as will be shown later other alicyclics show other relationships between substituents.

The general descriptions cisoid and transoid as applied to alicyclic stereoisomers are to be regarded as describing the configurational relationship of groups on a stereoisomer as drawn in the hypothetical 'vertical' Baeyer form. These terms are not applied to describe stereochemical relationships of groups on conformations since, in spite of the fortuitous equivalence found in cyclopentane derivatives, the terms need not be valid for modern perspectives as is clearly shown by cyclohexane derivatives. The interconvertible conformations (XLIX)–(LI) are some of the modern equivalents of the classical meso form but are not themselves definable as meso forms since they are dissymmetric. The modern perspectives having cisoid methyl groups on C_3 and C_4 are, however meso forms proper. The energy difference between the cisoid isomer (XLIX) and the transoid isomer (LII) is 1·7 kcal/mole; the latter conformation is the more stable (Haresnape[151]).

Molecular non-planarity also complicates the stereoisomerism of cyclohexane derivatives. The compound 1,2-dimethylcyclohexane has the three primary isomers represented by the Baeyer perspectives (LVII)–(LIX); the isomer (LVII) is the meso form and (LVIII)–(LIX) are enantiomers. The cyclohexane molecule exists in 'chair'

LVII   LVIII   LIX

and 'boat' forms; the permutation of the substituents on the chair conformation gives a total of six different perspectives namely (LX)–(LXV). These different perspectives, however, do not represent different primary isomers since internal rotation results in inverse puckering of the cyclohexane ring and, in turn, this results in the interconversion and equilibration of some of the isomers, thus, the biaxial form (LX) and the biequatorial form (LXIII) equilibrate. The pairs (LXI)–(LXIV) and (LXII)–(LXV) similarly equilibrate and the substance exists, therefore, as three primary isomer groups and six secondary isomers. The problem of correlating the stereochemistry of the classical and modern perspectives may be solved as described for 1,2-dimethylcyclopentane. Thus, the pairs (LX)–(LXIII) and (LXII)–(LXV) must represent enantiomers since, for example, (LX) and (LXV) are non-interconvertible mirror images; the pair (LXI)–(LXIV) are,

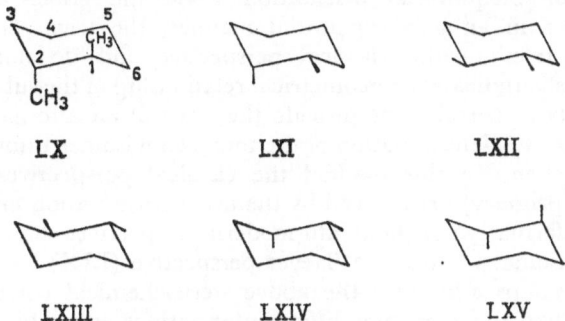

LX	LXI	LXII

LXIII	LXIV	LXV

therefore, the conformational equivalents of the classical meso form. In view of these relationships between the conformations and the hypothetical classical isomers the pairs of isomers (LX)–(LXIII) and (LXII)–(LXV) are described as transoid forms while the pair (LX)–(LXV) are the cisoid forms. The detailed correlation is now completed by considering the projected angle for the substituents. The projected angle has sign as well as magnitude and for cyclic molecules this is an important stereochemical datum and may be conveniently illustrated now for the case of cyclohexane derivatives. Thus, the molecule (LX) is arranged with the ring to the right of the C_1—C_2 bond (see p. 60), the Newman projection of this part of the molecule, looking along the C_1—C_2 bond and the angle determined. The projected angle is positive if measured anti-clockwise from C_1; the angle in (LXI) and (LXIII) is +60 degrees, in (LXII) and (LXIV) it is −60 degrees and in (LX) and (LXV) the angle is 180 degrees. The cisoid isomer pair (LXI–LXIV) have angles of ±60 degrees and so have the cisoid relationship of substituents found in aliphatic systems; the transoid pair (LX)–(LXIII) however, have their CH_3 groups in the aliphatic transoid (180 degrees) and cisoid (+60 degrees) relationships respectively. The isomer (LX) having a transoid classical configuration and a transoid orientation of groups on the modern perspective presents no nomenclature problem but (LXIII) is configuratively transoid yet on the conformation it has cisoid substitution. This latter isomer is, therefore, best referred to as the biequatorial-transoid or the skew-transoid form in order to define it. A further problem exists with the isomers (LXIII) and (LXI) since they have cisoid substituents of a different type, in the one case they are biequatorial and in the other case they are axial and equatorial. In summary a transoid 1,2-disubstituted cyclohexane has the substituents in a biaxial or biequatorial orientation while the cisoid isomer is substituted in an axial-equatorial manner, the terms cisoid and transoid refer to the classical perspective and do not, except fortuitously, indicate the geometrical relationship of the substituents.

The above correlations provide the basis of an alternative procedure for the determination of the total stereoisomer number for a cyclic system. In this method the classical perspectives for the isomers (primary) are derived by the usual permutation procedure; the transformation of these into modern perspectives then gives the possible isomers. Thus, the Baeyer perspective (LIX) has transoid methyl groups and from the above stereochemical correlation it follows that the corresponding conformations have biaxial substituents and biequatorial substituents. It should be remembered, of course, that the biaxial and biequatorial forms equilibrate and are related by internal rotation. The meso isomer gives an axial-equatorial conformation which by inverse puckering equilibrates with its equatorial-axial mirror image.

Cyclohexane stereoisomers exist also as 'boat' forms; these may be derived by general permutation using the 'boat' perspective or better they may be rotationally derived from 'chair' conformations. The isomers (LXVI)–(LXVIII) are some 'boat' forms derived from

LXVI   LXVII   LXVIII

(LXI); the 'boat' forms derived from a 'chair' form are easily interconvertible by mobile puckering of the ring and therefore, to the 'chair' secondary isomers must be added the collections of derived 'boat' isomers. The 'chair' conformations (LXI) and (LXIV), as an equilibrium mixture, constitute the optically inactive meso isomer of 1,2-dimethylcyclohexane. However, they are not meso forms themselves since they are enantiomeric but the equilibrium mixture is optically inactive. This example provides an interesting comparison of the conclusions of classical and modern stereochemistry.

The stability of the conformations of 1,2-dimethylcyclohexane follows the usual rules; 'boat forms' are highly unstable and can only be present in small amounts and of the 'chair' forms that having most equatorial groups is the most stable; thus the transoid biequatorial forms are the most stable conformations.

The above discussion is based solely on the operation of the steric factor; if two substituents are dipolar, for example halogen or carboxyl groups, then although the number and type of stereoisomers is the same as found for the 1,2-dimethylcyclohexane the stabilities of the isomers are different to those of the 1,2-dialkyl stereoisomers. Thus, for 1,2-dibromocyclohexane the biaxial conformation is the most stable because this orientation minimizes dipole repulsions.

Cyclohexanes having 1,3-substituents present a different case to the 1,2-compounds; the stereoisomers may . be determined as described previously and for 1,3-dimethylcyclohexane these are represented by the perspectives (LXIX)–(LXXIV). The inverse

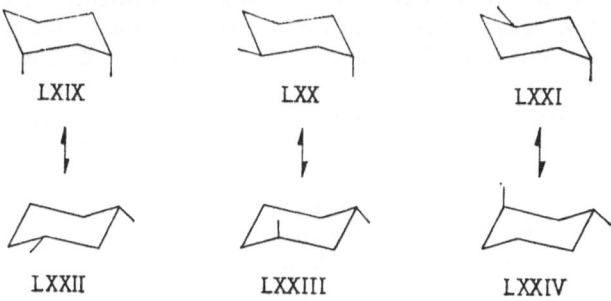

puckered equilibrated pair (LXIX)–(LXXII) are cisoid isomers because, like the classical cisoid perspective, they are symmetrical. The substituents in (LXIX) and (LXXII) also have the cisoid relationship, the projected angles being 0 degrees for each of the conformations. The transoid isomers (LXX)–(LXXIII) and (LXXI)–(LXXIV) have their substituents cisoid since projected angles are 60 degrees. In summary cisoid-1,3-disubstituted cyclohexanes have biaxial and biequatorial substitution whereas transoid isomers have the substituents in an axial-equatorial orientation.

It has been pointed out that the terms cisoid and transoid as applied to cyclic compounds derive from the classical period and refer to the relationship of groups on the 'vertical' Baeyer perspective;

the terms are now conventional and have no fundamental relevance to the actual stereochemistry of substituents on conformations. In the ethylenic series the *trans* isomer is usually most stable and this conclusion was applied to transoid cyclic isomers. This has caused some confusion because although the transoid forms of 1,2-dimethyl and 1,2-dibromocyclohexane are more stable than the cisoid form it is the cisoid isomer of 1,3-dimethylcyclohexane which is more stable than the transoid enantiomers. However, as pointed out, the difficulty is not real since the terms cisoid and transoid are applied as conventions and have geometrical meaning only when applied to the conformations.

The conformations (LXIX) and (LXXII) are the modern equivalent of the classical meso form of a 1,3-disubstituted cyclohexane; both conformations are symmetrical and so are meso forms themselves. Optical stereoisomerism is not shown by 1,4-disubstituted cyclohexanes because they have neither asymmetric centres nor general molecular dissymmetry; these compounds show only *cis-trans* isomerism.

Cyclohexane derivatives having three substituents have three asymmetric centres. This is of interest since a 1,4,x-tri-substituted compound illustrates the conversion of cyclic *cis-trans* isomers into optical isomers; thus 1,4-derivatives show only *cis-trans* isomerism but the introduction of a third substituent converts the *cis-trans* isomers to optical isomers. The determination of the stereoisomer number for polysubstituted cyclohexanes follows the previously discussed principles; in general, stability is associated with the largest number of equatorial substituents although other factors may operate to modify this conclusion.

The substance inositol (hexahydroxycyclohexane) is the classic example of a polysubstituted cyclohexane. The primary isomers of this substance (LXXV) are represented by the Baeyer perspectives (LXXXVI)–(LXXXIV). Inositol is of particular interest since inspection of the Kekulé-Couper structure (LXXV) does not reveal the presence of asymmetric centres and perspective permutation is necessary in order to show the optical stereoisomerism. This is, of course, necessary because the asymmetric centres present arise from a group geometry effect and not a constitutional factor. The particular stereoisomers (LXXVI) and (LXXXII) actually have no asymmetric centres but centres are present in all the other isomers; thus, (LXXIX) has four asymmetric centres and is a meso form; the symmetrical isomers (LXXVI) and (LXXXII) are not meso forms due to the absence of asymmetric centres. The isomers

(LXXXIII) and (LXXXIV) are enantiomers; they have six asymmetric centres. In summary, the classical analysis shows that inositol can have a pair of enantiomers, five meso forms and two symmetrical isomers. The Baeyer representations are converted to chair conformations by the application of the principles developed previously; thus, the hydroxyl group on C_1 of the Baeyer perspective (LXXVI) is introduced on to the conformation (LXXXV) in the

LXXV LXXVI LXXVII LXXVIII

LXXIX LXXX LXXXI

LXXXII LXXXIII LXXXIV

arbitrarily chosen axial position; the equatorial position could have been chosen equally well. The hydroxyl groups on C_1 and C_2 of the Baeyer diagram (LXXVI) are cisoid hence the hydroxyl group on C_2 of conformation (LXXXV) must be equatorial; in a similar way the complete stereochemistry of (LXXXV) can be derived. The conformation (LXXXV) equilibrates with (LXXXVI) but these are seen to be identical and therefore the Baeyer isomer (LXXVI) is represented by only one conformation. The corresponding conformations for the other Baeyer perspectives are shown as (LXXXVII)–(XCIV); each of these equilibrates, by inverse puckering, with another chair form; the exception is (XC) since

167

M

the two chair forms are identical. The conformations (LXXXVII)–(XCIV) have maximized equatorial substitution and therefore, when two chairs are possible the conformation drawn is the most stable. The isomer (XCII) is completely equatorial and is the most stable of the group; the isomer (LXXXV) is the least stable and

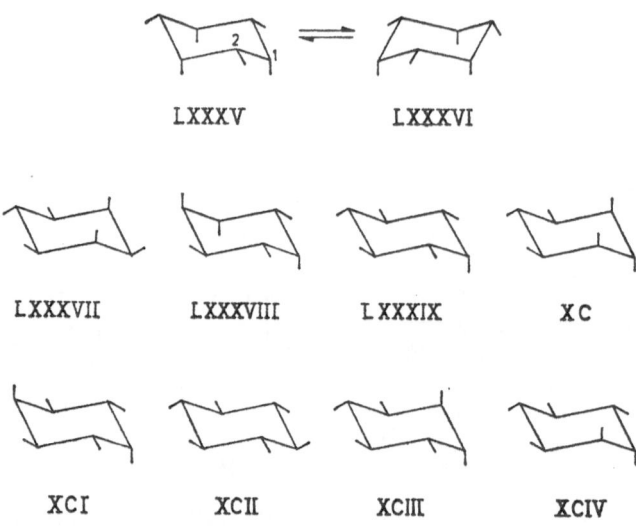

| LXXXV | LXXXVI |

| LXXXVII | LXXXVIII | LXXXIX | XC |

| XCI | XCII | XCIII | XCIV |

differs in free energy from (XCII) by 5·7 kcal/mole (Angyal and McHugh[151a]).

As mentioned earlier the stereochemistry of sugar molecules is closely similar to that of cyclohexane and of cyclopentane. The earlier researches in sugar chemistry revealed only the aliphatic structure of these substances and consequently their stereoisomerism was discussed in aliphatic terms (Fischer); thus, the D-glucose molecule (XCV) was represented by the Fischer projection (XCVI). Later work (Tanret[152]) showed that this isomer actually exists in two stereoisomeric forms which were called α- and β-D-glucose and this led to the postulation of cyclic structures. The cyclic structures for aldopentoses and aldohexoses are now well established; they exist as five and six membered rings containing an oxygen atom. The six membered (pyranose) ring structures are found in crystalline D-glucose (Mark[153], MacDonald and Beevers[154]); in solution some five membered (furanose) forms exist together with small amounts of the aliphatic form. In the classical period the α- and β-pyranose forms were accounted for and represented by the schemes (XCVI) → (XCVIII) and (XCVI) → (C). These schemes envisage a hydration of the aldehyde group i.e. —CHO → —CH(OH)$_2$ followed by the elimination of water between the hydroxyl groups on C$_1$ and C$_5$ in order to establish the ring. The projections (XCVIII) and (C)

clearly represent two stereoisomers and were satisfactory for classical purposes; however, these projections were drawn with the oxide ring 'on the right' without a proper analysis of its geometrical significance. The inaccuracies in these projections were removed by Drew and Haworth by considering the stereochemistry of ring formation; these authors also derived a Baeyer type per-spective as a more appropriate representation of the cyclic forms of sugar stereoisomers. Thus, the isomer (XCVII), for example, is drawn as the Fischer perspective (CI); the elimination of water between the hydroxyl groups on C$_1$ and C$_5$ requires them to be

169

as close as possible and the process occurs only with the molecule
in the form (CII). The cyclized molecule is represented by the
perspective (CIII) and by the projection (CIV); the oxygen ring
bonds do not appear in the projection since they lie directly below

| CI | CII | CIII | **CIV** |

the carbon-carbon bonds of the perspective and the oxygen atom
itself is best omitted for this reason.

Fischer perspectives for aliphatic systems do not represent
geometrical reality for two reasons (*a*) they are eclipsed structures
and (*b*) the carbon chain is represented as lying on a straight
line. This second property is, in fact, an advantage in aliphatic
stereochemistry but in order to convert a Fischer-Drew-Haworth
cyclic perspective such as (CIII) into the Baeyer type it must be

| CV | CVI |

constructed in eclipsed form from tetrahedral units. The construction of an eclipsed aliphatic molecule from tetrahedral carbons
gives a bent and not a straight chain and therefore in order to derive
the Baeyer perspective the structure (CII) must be drawn as (CV);
the need for the hydroxyl groups to come close together is clear
from the diagram (CV) and the cyclization process gives (CVI).

170

A further matter of stereochemical interest emerges from a consideration of the various perspective models for *D*-glucose. Thus *α-D*-glucose in aliphatic form may be represented as a Fischer or a modern perspective diagram but, of course, only the latter is geometrically real. The cyclic form of this stereoisomer is represented as (CVI) which may be regarded as geometrically real in respect of the general relationships between the substituents. This apparent anomaly of the derivation of a real perspective (CVI) from the unreal Fischer perspective is resolved when it is realized that the modern staggered perspective must be converted to the eclipsed form, by internal rotation, before it can be cyclized; the eclipsed modern perspective is, of course, identical to Fischer's bent perspective. Thus, although Fischer's aliphatic perspectives are geometrically unreal they have the advantage, in the sugar series, of being easily related to the cyclic forms of these molecules.

As indicated earlier, modern work has shown that six membered pyranose sugars have the regular skew hexagonal geometry characteristic of cyclohexane. The 'boat' and 'chair' forms are possible but for most sugars only the 'chair' forms exist. The 'chair' conformations of *α*- and *β-D*-glucose may be derived, as usual, from the Drew-Haworth perspectives and are shown, with maximum equatorial substitution, as (CVII) and (CVIII) respectively. The analogy with cyclohexane leads to the expectation that chair forms, for example (CVIII), will equilibrate with the inverse

CVII   CVIII   CIX

puckered form, for example (CIX), however, these alternative 'chair' forms are generally highly unstable and pyranoses exist a unique conformations.

PROJECTION DIAGRAMS FOR CYCLIC MOLECULES

Several forms of projection diagram may be used to represent cyclic stereoisomers and to provide an alternative method for the determination of stereoisomer number.

FISCHER-DREW-HAWORTH PROJECTIONS

The converse of the Drew-Haworth procedure converts a Baeyer perspective into a Fischer-Drew-Haworth (F–D–H) type of projection. These F–D–H projections are not, of course, projections of the molecule in Baeyer form but are derived projections. Thus, the Baeyer perspective of meso-cyclopropane dicarboxylic acid gives the

F–D–H projection (CX); in a similar way a 1,3-dibromocyclohexane gives the diagram (CXI). These projections are closely similar to aliphatic projections and permutation of substituents on the horizontals is the means of determining stereoisomer number.

NEWMAN PROJECTIONS

The use of Newman projections for alicyclic molecules involves nothing new; the use of the method is limited since only parts of a molecule may be represented satisfactorily in projection.

OTHER FORMS OF PROJECTION AND PERSPECTIVE DIAGRAM

A valuable partial perspective diagram is in general use particularly for describing polycyclic systems; these perspectives are drawings of the Baeyer models as seen from above. Thus, the primary isomers

of 1,2-dibromocyclohexane are represented by the diagrams (CXII—(CXIV); the full lines represent substituents lying above the ring plane and (CXII) is, therefore a cisoid isomer while (CXIII) and (CXIV) are the transoid forms.

Stereoisomers as conformations may be represented as partial projection–perspective diagrams. Thus, an appropriate arrangement of a 'chair' cyclohexane gives a planar hexagonal projection; the stereochemical relationship of the substituents is then indicated on this projection by the use of full and broken lines. The isomers (CXV)–(CXVII) are represented in partial projection-perspective

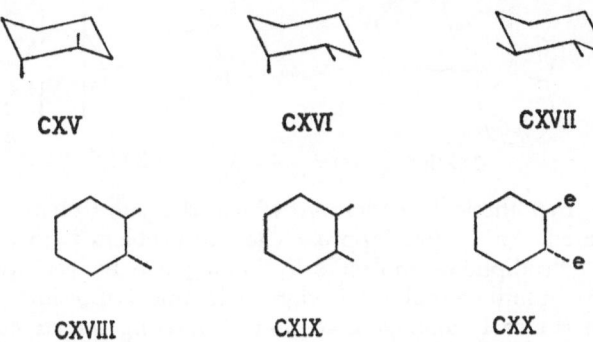

CXV   CXVI   CXVII

CXVIII   CXIX   CXX

by (CXVIII)–(CXX) respectively. The use of full and broken lines in this latter type of diagram refers only to the relative configuration of substituents; the isomers may be defined completely by introducing the a and e symbols into the diagram as shown in

CXXI   CXXII

(CXX). A simplified version of these diagrams, due to Linstead[155], uses a filled circle to represent the above plane substituents thus, the cisoid isomer (CXII) or (CXIX) and the transoid form (CXIII) or (CXX) are represented by (CXXI) and (CXXII) respectively.

POLYCYCLIC MOLECULES

NON-FUSED MOLECULES

Dodecahydrodiphenyl (perhydrodiphenyl) derivatives are examples of this type. The parent hydrocarbon may be regarded as a monosubstituted cyclohexane and has no asymmetric centres. However, the compound may exist as an equilibrium mixture of the three conformations $a:a$, $a:e$ and $e:e$ with the more stable $e:e$ form predominating in the mixture. The introduction of a substituent, as in 2-bromoperhydrodiphenyl, creates two different asymmetric centres if the analogy with cyclohexane is assumed. The 2^n-rule predicts four stereoisomers for the compound but, in the absence of actual geometrical knowledge, their representation requires a

CXXIII   CXXIV

suitable hypothetical model on which the substituents can be permutated. An obvious approach to the problem is to construct the perhydrodiphenyl molecule by joining two Baeyer cyclohexyl units. An infinite number of ways of joining is possible and the problem is clearly analogous to that of deriving the stereoisomers of tartaric acid starting with two tetrahedral units. The choice of mode of joining of the cyclohexyl groups is arbitrary but by analogy with ethane the staggered form (CXXIII) is a sensible selection; this perspective may be simplified to (CXXIV) as was done for cyclohexane and is justified for the same reasons. The permutation of the 2-bromine atom on (CXXIV) gives the four possible stereoisomers represented by (CXXV)–(CXXVIII). The described

CXXV   Br   CXXVI   Br

CXXVII   CXXVIII

method is a derivation of stereoisomer number from first principles; alternatively the simpler method of regarding the compound as a substituted cyclohexane may be used. Thus, by considering each ring in turn as a substituted cyclohexane the compound perhydro-diphenic acid gives the isomers (CXXIX)–(CXXXVIII); the isomers are made up of two meso forms (CXXIX) and (CXXX) together with four pairs of enantiomers. This method of deriving the isomers, contrary to the previous method, shows some isomers as having cisoid as well as transoid bridge hydrogens. This is a defect of this second method since a transoid relationship is geometrically more correct. Thus, this second method is analogous

COOH COOH

CXXIX

CXXX

COOH COOH

CXXXI

CXXXII

COOH COOH
COOH COOH

COOH ·COOH
COOH COOH

CXXXV

CXXXVI

CXXXIII

CXXXIV

COOH COOH
COOH COOH

COOH COOH
COOH COOH

CXXXVII

CXXXVIII

to a representation of meso-tartaric acid in the eclipsed form and an enantiomer in the staggered form. The above stereoisomers may be represented, of course, as conformations and the stable $e:e$ perspective is used, this perspective has its bridge hydrogens in the geometrically correct transoid form.

FUSED MOLECULES

A variety of polycyclic molecules is known; the type derived by fusion through two adjacent carbon atoms is the most important one and occurs in many natural products. The compounds decahydronaphthalene (decalin) (CXXXIX) and hydrindane (CXL) are simple examples of these 1,2-fused polycyclics.

The Detection of Asymmetric Centres

A 1,2-fused polycyclic molecule, unlike the non-fused type such as perhydrodiphenyl, may or may not have the fusion atoms as asymmetric centres; thus (CXXXIX) has no asymmetric centres but both fusion carbon atoms in (CXL) are asymmetric ones. The

CXXXIX   CXL

detection of asymmetric centres in fused molecules follows the method described for monocyclics. The asymmetric centres in monocyclics carry only two exacyclic substituents and so the two paths round the molecule correspond to the other two substituents; these paths are different if the centre is asymmetric. Thus, in (CXL) and considering the centre C_9 all the paths round the molecule are different and so the centre is asymmetric; the procedure applied to the fusion atoms of (CXXXIX) shows that these centres are not asymmetric. A general rule for the detection of asymmetric centres in fused polycyclics may be deduced, namely that if the two fused rings are of different size then the fusion atoms are asymmetric.

CXLI   CXLII

The decalin molecule has no asymmetric centres but is capable of *cis-trans* isomerism and, in accordance with the general principle found for monocyclics, the introduction of a substituent at a position other than a fusion atom creates three asymmetric centres and optical stereoisomerism. The molecule (CXLI) has no asymmetric centres but (CXLII) has three centres.

The Decalins and Hydrindanes

Classical theory—The hydrindanes can be constructed as planar Baeyer type models made up of a regular hexagon and a regular pentagon. The hydrogen atoms attached to the fusion carbons must be orientated, in relation to the three carbon-carbon bonds, as near tetrahedral as possible in order to minimize strain; this may be done in two ways and results in the hydrogen atoms being orientated cisoid and transoid. These models are, of course, highly strained and their existence was doubted by classical theory, however, they have

CXLIII **CXLIV** **CXLV**

the validity that they correctly predict the stereoisomer number. Thus, these models give the possible isomers of hydrindane as (CXLIII)–(CXLV); the pair (CXLIII)–(CXLIV) are enantiomers and (CXLV) is a meso form as required by the presence of two similar asymmetric centres.

Decalin can exist only as *cis* and *trans* isomers but 2-decalol has the isomers (CXLVI)–(CXLIX) derived by the permutation of the

CXLVI **CXLVII**

CXLVIII **CXLIX**

hydroxyl group on the *cis* and *trans* decalins; each of the isomers is dissymmetric and therefore, a total of eight stereoisomers exists in accordance with the 2^n rule.

Modern theory—Primary *cis-trans* isomerism in decalin has been shown experimentally to be real (Huckel[156]); conformational theory shows their geometries to be (CL) and (CLI) respectively. These models differ in an important way from the classical models since the rings are fused *cis* and *trans* as well as having the hydrogens, attached to the fusion carbons, *cis* and *trans*; the classical models have only the hydrogen atoms in a *cis* and *trans* relationship. The determination of stereoisomer number, in modern terms, permutates substituents on the perspectives (CL) and (CLI). Thus, for 2-decalol and considering only the *trans* perspective the permutations are

CL   CLI

(CLII)–(CLV); the pairs (CLII)–(CLIV) and (CLIII)–(CLV) are enantiomers. The analysis must be completed by considering internal rotation, however, *trans*-decalin differs from cyclohexane in that it is a rigid structure, the usual inverse puckering does not occur, and therefore the perspectives (CLII)–(CLV) represent primary isomers.

CLII   CLIII

CLIV   CLV

Substituted *cis*-decalins present a more complex problem since these are flexible molecules. The permutation of the hydroxyl group gives four different perspectives two of which are (CLVI) and (CLVII). The isomer (CLVI) equilibrates with (CLIX) by passing through the 'boat' form (CLVIII); the stereochemical relations are seen best by drawing (CLIX) as (CLX). The isomer (CLVII) similarly equilibrates with (CLXI); thus, in spite of internal rotation the axial isomer (CLVI) is not converted into the equatorial form (CLVII). The *cis* 2-decalol, therefore, exists as two pairs of primary enantiomers and each of these equilibrates with a secondary form.

An aliphatic molecule having n different asymmetric centres exists as 2^n stereoisomers, however this does not always apply to

CLVI

CLVII   CLVIII

CLXI   CLX

CLIX   CLXII

polycyclic systems. The compound dihydrocholesterol (CLXII) has nine different centres and is capable, theoretically, of having 512 stereoisomers; some strainless conformations can be constructed but the majority of isomers are highly strained and cannot exist.

COMBINATIONS OF ALIPHATIC AND ALICYCLIC SYSTEMS

The substance (CLXIII) is an example;

CLXIII

CLXIV

it may be regarded as a substituted cyclopentane (CLXIV) and as such exists as four isomers; each of these can have the enantiomeric forms of R attached making a total of eight isomers in agreement with the 2^n rule.

AROMATIC MOLECULES

Aliphatic and alicyclic molecules containing asymmetric centres have been discussed above; aromatic molecules do not contain asymmetric atoms in the nucleus and so do not show optical isomerism of this type.

8

OPTICAL STEREOISOMERISM: (IV) MOLECULES HAVING GENERAL MOLECULAR DISSYMMETRY

INTRODUCTION

An aliphatic or alicyclic tetrahedral atom of carbon, nitrogen etc. is asymmetric if no plane of symmetry cuts it and two of the substituents and passes between the remaining two substituents; this definition must be modified for other atoms such as p^3N since the test planes now cut the nitrogen atom and one substituent and pass between the other substituents. This analysis may be extended to other atoms such as sp^2 carbon although the present practice does not consider this type of atom; thus by accepted definition asymmetric centres refer to non-multiply bonded aliphatic and alicyclic atoms though it must be emphasized that there is no fundamental reason for excluding other species. Asymmetric centres which depend on the constitutional differences of their substituents are detected simply by inspecting the Kekulé-Couper structure of the compound. However, as the trihydroxyglutaric acid problem shows, the simple inspection method of detection does not necessarily suffice, to deduce the total isomer number since the group geometry effect may operate, centres which are asymmetric because of the group geometry effect can be detected only when a hypothesized or actual geometrical model is available. Thus, the detection of centres of this type is possible only when the stereoisomers have been determined by other means, i.e. by permutation, although this indirect, detection of centres is useful for purposes of classification. Many molecules having no asymmetric centres show optical isomerism because their inherent geometry is dissymmetric, again, the detection of this form of dissymmetry requires hypothesized or actual molecular models. The total number of stereoisomers for a substance can be derived, unambiguously, by the permutation method, with or without the use of projection diagrams and two or more stereoisomers may exist. If only two dissymmetric isomers exist and no asymmetric centres are present then the molecules show general molecular dissymmetry. When a group of more than two isomers exists then, if the 'four different groups' rule is employed, all or none of the isomers will contain asymmetric centres but if the group geometry test is used all, none or some of the isomers will contain asymmetric centres. It will be clear that the definition of isomers as showing general molecular dissymmetry depends on the definition of asymmetric centres used. Historically, optical isomerism due to general molecular dissymmetry has been predicted on the basis of an existing molecular model, for example the allenes (van't Hoff) and, conversely, geometrical models have been formulated as a consequence of the detection of isomerism, for example the diphenyls (Turner and Lefevre[157]; Bell and Kenyon[158], Mills[159]). The various types of compound showing general molecular dissymmetry may now be discussed.

ALIPHATIC MOLECULES

Ethylene dibromide exists as a symmetrical staggered and two dissymmetric skew forms. The 'four different groups' rule shows that no asymmetric centres are present and the skew forms are regarded as showing general molecular dissymmetry. However, invoking the group geometry effect shows that the skew forms do contain asymmetric centres although the staggered form does not and therefore the skew isomers would not be regarded as showing general molecular dissymmetry. The ethylene dibromide isomers were described earlier as *trans* and *cis* but the above discussion shows that the skew (*cis*) forms are optical isomers of the asymmetric centred type or of the general molecular dissymmetry type.

Ethylenic units are associated usually with *cis-trans* isomerism but some ethylenic systems are dissymmetric and optical isomers exist. Thus, large *cis* substituents can overcrowd intramolecularly so bending the substituent bonds out of the ethylenic plane and/or twist the double bond so producing dissymmetry. These two non-planar dissymmetric *cis* isomers are clearly a case of general molecular dissymmetry since no asymmetric centres exist. Optical isomers of this type are interconvertible easily by a vibrational process and so are secondary enantiomers. The allenes are an historically important example of general molecular dissymmetry created by ethylenic units. Optical isomerism was predicted for allenes of the type (I) (van't Hoff) since a model based on classical ethylenic carbon shows the substituents lying in planes at right angles to each other so making the molecules dissymmetric. The quantum mechanics and experiment have since confirmed van't

Hoff's general geometry for these compounds, thus the allene (II) has been obtained as primary enantiomers (Mills and Maitland[160]). The synthesis of allenes has proved rather difficult but some earlier support for the van't Hoff geometry was obtained in 1909 (Pope, Perkin and Wallach[161]) by the isolation of enantiomers of the allene

I   II

analogue (III). The chain extended allenes are called cumulenes; butratriene (IV) (Kuhn and Wallenfels[162]) is the simplest member. The end substituents in (IV) lie in the same plane and optical isomerism is impossible, however, in appropriate derivatives *cis-trans* isomerism can exist. The five carbon cumulene (V) and other

III   IV

$$CH_2=C=C=C=CH_2$$

V

odd membered cumulenes are geometrically similar to allene and can show primary optical isomerism due to general molecular dissymmetry.

ALICYCLIC MOLECULES

Polycyclic molecules derived by the fusion of two alicyclic rings through a common sp^3 carbon atom as in (VI), constitute parent molecules which may be appropriately substituted to give dissymmetric geometries, for example (VII) (Jansen and Pope[163]), is capable of primary optical isomerism. The classical geometry of these molecules was based on the distorted Baeyer tetrahedral unit but the symmetry properties are the same as those based on modern geometrical data although the cyclobutane rings presumably have some distortion. These compounds are called spiro-compounds and heteroatom analogues such as (VIII) have been obtained as primary enantiomers; on the basis of the 'four different groups' rule these

N

spiro-compounds do not contain asymmetric centres and therefore show general molecular dissymmetry. However, when group geometry is invoked the isomers obviously have asymmetric centres. The alicyclic species (IX) is an unambiguous example of general

VI

VII

VIII

IX

molecular dissymmetry; it has no asymmetric centres and the iodine atoms experience non-bonded interaction giving a dissymmetric molecule.

AROMATIC MOLECULES

INTRODUCTION

Classical theory envisaged benzenoid and heterocyclic aromatic molecules as planar symmetrical structures and therefore the discovery of optical isomerism in this group was of great stereochemical importance (Christie and Kenner[164]).

Aromatic optical isomerism is due to general molecular dissymmetry since these systems cannot have asymmetric centres as parts of the rings. The dissymmetry of aromatic systems arises in two general ways namely (a) due to the existence of intramolecular barriers (steric and/or conjugational) and (b) due to the distortion of a normally symmetrical aromatic system by steric overcrowding. Thus, the principles governing optical isomerism in aromatic compounds are the same as those previously found applicable to certain aliphatic compounds. Historically, aromatic optical isomers of the barrier type were discovered before the analogous aliphatic type, such as the skew ethylene dibromides, because aromatic units happen to be very suitable geometrical devices for the creation of high internal barriers. Thus, aromatic isomers having high barriers to interconversion are primary and are easily detected and isolated by classical techniques. Aliphatic molecules have low barriers and the detection of optical isomers requires modern methods.

Primary aromatic optical isomers also result from overcrowding and again aromatic species are suited particularly for the manifestation of this effect whereas analogous aliphatic types, such as *cis*-ethylenes give only secondary optical isomers. Aromatic isomers arising from overcrowding present difficulties in synthesis and so are rarer than the rotationally hindered type and indeed have been obtained only recently (Newman and Hussey[165]).

MOLECULES HAVING INTERNAL BARRIERS TO ROTATION

The Diphenyls

Introduction—As discussed earlier the diphenyl molecule has two internal barriers to the rotation of the rings about the pivot bond. Thus, the non-bonded repulsion of the hydrogen atoms in the 2,2'- and 6,6'-positions creates a steric barrier and further, the molecule is conjugated and gives the pivot bond some double bond character which creates a resonance barrier. Steric interaction (destabilization) is maximum for a planar molecule (barrier height 5 kcal/mole); resonance is at a minimum (destabilization) at a rotation of one ring through 90 degrees from the coplanar position (barrier height c. 3 kcal/mole). These two effects equilibrate so that the gaseous or solution phase has non-planar molecules and the rings are rotated to 45 ± 10 degrees from the coplanar position. In the crystal the environment factor enforces coplanarity of the molecules.

A non-planar diphenyl molecule (X) is clearly dissymmetric and exists in enantiomeric form; the enantiomers may be represented as the projection diagrams (XI) and (XII). These projections

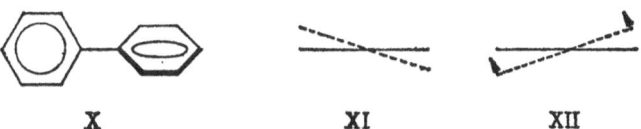

X   XI   XII

correspond to an end-on view of the molecule; the full line represents the ring nearest to the eye. The enantiomers can be interconverted by a rotation of the distant ring of, say (XII), through either the steric barrier (clockwise) or through the resonance barrier (anticlockwise) as indicated by the arrows. In diphenyl itself these barriers are low and the interconversion of enantiomers at ordinary temperatures will occur by both paths. Thus, diphenyl in the gaseous or solution state is an optically inactive equilibrium mixture of the enantiomers; crystalline diphenyl is optically inactive because it consists of planar symmetrical molecules. The transition state for a steric barrier interconversion is, of course, the planar molecule; an interconversion via the resonance barrier gives a transition state with the rings at right angles.

The introduction of substituents into diphenyl may increase the heights of the internal barriers; the steric barrier can be very considerably raised by substitution whereas the resonance barrier cannot be changed greatly. The steric barrier is affected most by ortho substituents and naturally diphenyls of this type are of greater stereochemical interest. The height of the steric barrier, i.e. the activation energy, determines for appropriate diphenyls whether, at a particular temperature, they exist as primary or secondary enantiomers. Historically the recognition of the existence and function of the steric barrier is due to Turner and Lefevre[157], Bell and Kenyon[158] and Mills[159]; the low resonance barrier can create only secondary isomers and the recognition of this type of barrier is, of course, a development of quantum theory. The dependence of the height of the steric barrier on the size of the substituents, particularly ortho substituents, results in diphenyls of all degrees of stability as has been shown by the extensive experimental studies of Adams[165a].

Monosubstituted diphenyls—A substituent may be orientated in the ortho, meta or para position in a ring of the diphenyl molecule;

as pointed out ortho substitution is most relevant to the problem of optical isomerism. The substance 2-bromodiphenyl (XIII) has a steric barrier made up of the two non-bonded interactions Br · · · · H and H · · · · H, this steric barrier is clearly higher than in diphenyl itself although the resonance barrier is essentially unchanged. In the gaseous and solution state and also probably in the crystal the molecules of (XIII) are non-planar and the enantiomers are represented by the projections (XIV)–(XV); this latter projection may be drawn alternatively as (XVI) and clearly (XIV) may be converted to (XVI) by a rotation of the unsubstituted ring through the low resonance barrier. One concludes that mono-ortho substituted diphenyls, independent of the height of the steric barrier can show only secondary isomerism since the enantiomers can be interconverted without the need of rotation through the steric barrier. The above analysis is incomplete, a determination of the total number of stereoisomers may be obtained by the usual permutation. Thus, the permutation of the bromine atom on each of the enantiomeric parent diphenyl molecules gives eight permutations, however, only two of these are different and they correspond to the enantiomers (XIV) and (XV).

The effect of meta or para substituents is small but in a detailed analysis it cannot be ignored. Thus, a meta substituent may buttress the adjacent ortho hydrogen atom bending it towards the ortho hydrogen on the other ring and so slightly increasing the steric barrier. A para substituent may have an effect, though this is not

XVII

XVIII

XIX

XX

completely understood. These effects are small and so diphenyls having only meta or para substituents show only secondary isomerism.

Disubstituted diphenyls—The molecules (XVII)–(XX) are some generalized examples of this type. The permutation of the groups A and B, as in (XVII), on the two diphenyl parent molecules gives sixteen permutations but of these only the four (XXI)–(XXIV) are different and clearly these consist of enantiomer pairs (XXI)– (XXII) and (XXIII)–(XXIV). The interconversion of (XXI)

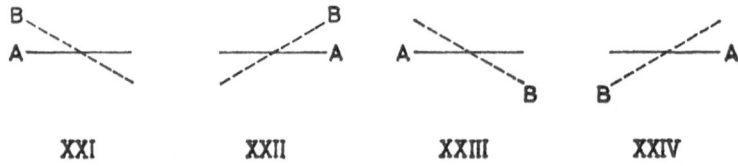

| XXI | XXII | XXIII | XXIV |

and (XXII) is best seen by drawing out the projections as (XXV) and (XXVI) respectively; it follows that (XXV) can be converted into (XXVI) by rotation through a steric barrier however, two paths of rotation are possible. Thus, group B may squeeze past A by an anticlockwise rotation of the ring containing B or B may

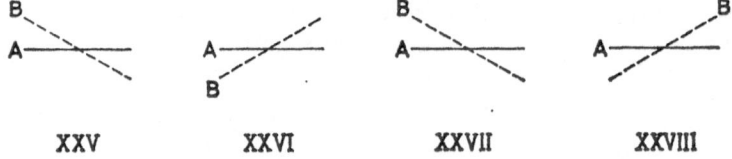

| XXV | XXVI | XXVII | XXVIII |

squeeze past the hydrogen of the A ring by a clockwise rotation. The steric barriers for these rotations are made up of $B \cdots A$ and $H \cdots H$ interactions for anticlockwise motion and for the clockwise path $A \cdots H$ and $B \cdots H$ interactions occur; the latter path involves the smallest repulsion and is preferred. The interconversion of (XXIII) and (XXIV) similarly takes place by rotation through a steric barrier. However, it is seen when (XXI) and (XXIV) are drawn in the alternative forms (XXVII) and (XXVIII) that they are interconvertible by a rotation through the resonance barrier. Thus, the diphenyl (XVII) exists as two primary enantiomers each made up of the pairs of secondary forms (XXI)– (XXIV) and (XXII)–(XXIII). The conformations (XXI) and (XXIV), for example, are also different in energy; physical evidence shows (Bastiansen[166]) that the conformation (XXI), in which the substituents are closest together is most stable because van der Waals attraction between A and B exceeds that of either A or B with a hydrogen atom. The reality of van der Waals attraction is shown by 2,2′-dichloro-, 2,2′-dibromo, 2,2′-di-iodo-, 2,2′-dinitro- and 2,2′-dicarboxydiphenyl; thus considering the angle of twist to be zero when the substituents are eclipsed the twists in the above molecules are found to be 74 degrees (or 62 degrees), 75, 79, 90 and 90 degrees respectively. The secondary isomers of (XVII) are present only in the gas or in solution, in the crystal only two isomers will persist. One concludes that a 2,2'-disubstituted diphenyl having sufficiently large substituents is capable of primary stereoisomerism and that in an appropriate state of aggregation each enantiomer equilibrates with a secondary form. The diphenyls (XXIX) (Lesslie and Turner[167]) and (XXX) (Shaw and Turner[168]) are examples of diphenyls of this type.

XXIX   XXX   XXXI

The substance (XVIII) could have a high steric barrier created by the interaction of A with hydrogen on the other ring, however, as projection diagrams show, the group B is so orientated that enantiomers are interconverted readily by a rotation through the resonance barrier and the compound can exist only as secondary isomers in spite of the high steric barrier. One concludes that primary isomerism in diphenyls depends on (a) the existence of a sufficiently high steric barrier and on (b) the appropriate orientation of substituents. These conclusions are exemplified further by (XIX) and (XX), the former has its enantiomers easily interconverted through the resonance barrier while the latter is capable of primary isomerism if A is large enough; the molecule (XXXI) is an example of this latter type and has been obtained as primary enantiomers (Lesslie and Turner[169]).

Polysubstituted diphenyls—A large number of optical isomers of this kind is known; the compounds (XXXII)–(XXXIV) are examples. The projection diagrams for the primary enantiomers of (XXXII) are drawn as (XXXV)–(XXXVI) since the carboxyl groups will be closest in the most stable conformations. Thus, (XXXII) is capable of potential primary isomerism since enantiomer interconversion must proceed through a steric barrier. In a similar way (XXXIII) is potentially capable of primary isomerism but (XXXIV) is not. The detection of potential primary isomerism in diphenyls is done most fundamentally by the projection diagram method described. However, as an alternative a simple rule may be applied. Thus, the diphenyl molecule is drawn in the plane of the paper, for example (XXXII), one aromatic ring, for example that containing the NO_2 group is now imagined rotated through

XXXII   XXXIII   XXXIV

90 degrees, if the molecule so orientated has no plane of symmetry then primary isomers are potentially possible, experiment shows that (XXXIII) exists as primary enantiomers but the steric barrier in (XXXII) is not high enough for primary forms.

XXXV   XXXVI

Further discussion of the steric, energetic and environment factors—As shown, a substituted diphenyl potentially capable of primary optical isomerism will actually show isomerism if the steric barrier is high enough. The presence of a high steric barrier in a diphenyl may be inferred, of course, from the experimental isolation of enantiomers but, clearly, it is desirable to have a theoretical means of predicting primary isomerism for any particular diphenyl.

An earlier approach to this problem took the form of an empirical rule (Stanley and Adams[170]); the rule states that if in a diphenyl (XXXVII) the sum of the bond lengths $d_1 + d_2 \geqslant d_3$ (2·9 Å) then a barrier exists sufficient to create primary isomers. The application of the rule to (XXXVII) assumes that rotation occurs through the steric barrier $A \cdots C$ and $B \cdots D$ and that of these two inter-actions the one $A \cdots C$ is the larger in the planar molecule. This empirical rule also regarded the excess of the sum $d_1 + d_2$ over

2·9 Å as a semi-quantitative measure of the ease of interconversion of enantiomers. The rule is very satisfactory for 2,2′,6,6′-tetra-substituted diphenyls but, in general, not for 2,2′,6′- tri- or 2,2′-disubstituted molecules.

XXXVII   XXXVIII

A second, theoretical, approach to the prediction of primary isomerism of diphenyls is due to Mills and Lesslie and Turner[170a]. This method uses models or scale drawings of molecules in order to determine the degree of mechanical interaction of ortho sub-stituents; the atomic radii used were derived from the available x-ray data. This method also proved useful only in the case of 2,2′, 6,6′-tetra-substituted systems.

These two early methods use covalent radii as the steric basis for determining non-bonded interaction but as shown earlier the proper criterion for steric interaction is the van der Waals radius or a datum related to it; further, other factors operate to reinforce (or otherwise) an internal steric barrier. A detailed quantitative analysis of the diphenyl problem has become possible only recently as a result of the development of modern theory and the availability of accurate physical data; such an analysis has been given for the diphenyl (XXXVIII) by Hill[171] and by Westheimer and Mayer[172].

The earlier theories assumed mechanically rigid molecules but on this basis the non-bonded interaction in the planar form of (XXXVIII) is 200 kcal/mole. This activation energy is clearly prohibitively high and the enantiomers of (XXXVIII) would decompose rather than be interconverted. Experiment shows that in fact, (XXXVIII) does exist as primary isomers but the activation energy is found to be only about 17 kcal/mole. Most primary enantiomeric diphenyls have activation energies in the range 19–23 kcal/mole although 2,2′-,6,6′-derivatives can have energies as high as 40 kcal/mole; a diphenyl having an activation energy of less than 17 kcal/mole cannot show primary isomerism at room temperature. Modern theory shows that molecules are not rigid structures and that non-bonded interaction can be relieved by the bending, stretching or compression of bonds as well as by other secondary forms of distortion. This molecular elasticity results in a much smaller activation energy than for a rigid molecule; bonds are relatively easy to bend though more difficult to stretch or compress and calculation using the elastic model gives a value of 18·2 kcal/mole for the activation energy of (XXXVIII).

As pointed out earlier, ortho substituents are more directly important to diphenyl isomerism but meta and para substituents can also affect the stability of enantiomers. Thus, the diphenyl (XXXIX) has a meta substituent and its enantiomers are more

XXXIX   XL

XLI

stable than those of (XL) in spite of the fact that the ortho substitution is the same in both cases. The buttressing effect interprets this increased stability, thus, the 3'-NO_2 group in (XXXIX) buttresses the adjacent —OCH_3 group making it more difficult for this latter group to be bent back in the formation of the transition state. The operation of this effect increases the activation energy. The enantiomers of (XLI) are more stable than the corresponding molecules lacking the 4'-NO_2 group; the stabilizing effect of such groups was attributed earlier to their mesomeric effect (e.g. Calvin[173]). Thus, the operation of this effect would give the pivot bond partial double bond character and the shortened bond will increase the ortho steric effects. However, consider a planar diphenyl system with its attendant steric effects and suppose, now, that a resonance effect is introduced, it is clear that the resonance effect will actually operate only if it leads to an increased stability. The data show that the activation energy actually increases hence in the planar form of (XLI) although the orbitals of both rings are parallel there is effectively no overlap between the rings since the steric factor prevents the pivot bond from contracting. The function of para groups in molecules such as (XLI) is not wholly understood.

The detailed consideration of the energetics of the interconversion of diphenyls involves both the energy and the entropy of activation. The entropy factor appears to be of general importance in the diphenyl series since the rates of interconversion of enantiomers vary

XLII   · XLIII

considerably but activation energies vary little and are generally in the range 19–23 kcal/mole (Cagle and Eyring[174]). This variation in rate clearly must be due to variation in the entropy of activation. Thus, the interconversion of the enantiomers of (XLII) and XLIII) involves the same activation energy of 22·6 kcal/mole (Ahmed and Hall[175]) and yet their interconversion rates are quite different; these different rates are entirely due to the different entropies of activation.

XLIV   XLV

XLVI   XLVII

Ring bridged diphenyls—The phenanthrene molecule (XLIV) may be regarded as a diphenyl molecule linked through two ortho positions by an ethylenic unit; the stereochemistry of an ethylenic unit together with resonance gives a planar molecule. The fluorene molecule (XLV) is expected to be planar, as in fact it is, but the shortness of the bridge results in some bending of the pivot bond and some deformation of the tetrahedral angle of the CH_2 group. The dihydrophenanthrene (XLVI) if planar has two sources of instability namely the interaction of the pair of ortho hydrogen atoms and the interaction of the eclipsed hydrogens of the ethane bridge. Thus, a rotation of the rings about the pivot bond is expected until the relief of the steric interactions equilibrates with the deformation of bond angles or other distortions in the bridge structure. The steric barrier in an enantiomer of (XLVI) is roughly equal to the sum of two ortho diphenyl interactions and the eclipsed ethane interactions; this sum is low hence the molecule gives only secondary isomers. Primary isomers can be derived from (XLVI) by using larger groups in the unsubstituted ortho positions. The conclusions for (XLVI) also apply to (XLVII).

The diphenyl (XLVIII) is of historical interest in connection with the origin of isomerism in the diphenyl series. This molecule, which is optically inactive, is synthesized by cyclizing the primary enantiomer (XLIX); the molecule (XLVIII) is expected to be

XLVIII   XLIX

planar since the —NH·CO— unit can have a planar *cis* form. The inactivity of (XLVIII) and the inferred planarity was considered to be evidence that the enantiomer (XLIX) owes its activity and isomerism to non-planarity.

L   LI

Other Diaryls

A variety of other diaryls showing optical isomerism is now known; their isomerism follows the principles discussed for diphenyls. The compounds (L) and (LI) are examples of diaryls which have been obtained as primary enantiomers. The primary isomerism

194

of dipyridyl (LI) is surprising, perhaps since it has no ortho groups, however, the nitrogen atoms have lone pairs and these are presumably responsible for the steric barrier.

Other Aromatic Molecules

The steric barrier theory of the origin of isomerism was extended to other molecules containing aromatic units by Mills and Elliott[116]. Thus, models suggested that molecules of the type (LII) have a steric barrier to rotation about the C—N bond due to interaction between NO_2 and R' or R''. The analogy with the diphenyls is complete because (LII) also has a low resonance barrier due to the

LII LIII LIV

conjugation of the lone pair on the nitrogen atom with the ring. The expectation was verified when (LIII) was obtained as primary enantiomers; the theory was further supported by the failure of the unhindered (LIV) to show primary isomerism.

LV LVI

A large number of aromatic compounds of this general type is now known; the compounds (LV) and (LVI) are further examples. The analogue of (LV) lacking the bromine atom, can be converted into its enantiomer without passing through the steric barrier; this is possible because of the form of the substitution and it emphasizes the analogy to the diphenyls. However, in spite of the general analogy between diphenyls and these other aromatic types some differences of detail exist, thus the effect of substituents on the steric barrier is different in the two series because of the different geometries.

MOLECULES HAVING INTRAMOLECULAR OVERCROWDING

Molecules within which non-bonded repulsive interactions exist and which cannot be relieved without creating some further form of intramolecular steric strain are said to be overcrowded. The relief of intramolecular repulsive strain in an overcrowded species results in various changes in molecular geometry and symmetrical molecules may be made dissymmetric.

The eclipsed ethane molecule is overcrowded but since the non-bonded interaction may be relieved by staggering and without inducing other forms of strain the ethane molecule is not defined as overcrowded proper. A planar diphenyl molecule is overcrowded and its relief is possible only by the twisting of the partially double

pivot bond. However, the isomerism of the diphenyls does not depend on this resonance strain and these compounds are best considered as an example of restricted rotation rather than over-crowding.

The incidence of intramolecular overcrowding is widespread but usually enantiomers so created are easily interconverted, at ordinary temperatures, by vibration. However, some aromatic species have large overcrowding effects and primary enantiomers may be isolated by classical methods.

The compounds (LVII), (LVIII) (Newman and Hussey[165]), (LIX) (Bell and Waring[99]) and (LX) (Theilaker and Baxmann[177]) have been obtained as enantiomers of rather low stability, they are interconverted fairly readily at room temperature. The compound

LXI

LXII

(LXI) (Newman[178]) gives highly stable enantiomers but (LXII) (Wittig and Zimmerman[179]) is only weakly overcrowded and does not give primary enantiomers.

MOLECULES HAVING ASYMMETRIC CENTRES AND GENERAL MOLECULAR DISSYMMETRY

Compounds of this type are not difficult to synthesize but they happen to be rare; the substances (LXIII) (Robinson[180]) and (LXIV) (Wittig[179]) are examples and are cases of ring bridged

LXIII

LXIV

diphenyls. A diphenyl can exist as two primary enantiomers and hence a diphenyl substituted by a classical asymmetric centre can exist in a total of four forms. The number of isomers may be deduced also from models, however, these methods are open to doubt when applied to complex systems and an unambiguous analysis requires actual geometrical knowledge. A reasonable deduction of the isomer number for (LXIV) on the basis of fundamental stereochemical

LXV   LXVI

principles may be made, thus starting with a planar molecule, the relief of non-bonded interaction can give the dissymmetric species (LXV) and (LXVI). These systems can exist as enantiomers making a total of four forms. Alternatively, having decided on a reasonable perspective for the parent hydrocarbon, group permutation gives the isomers.

9

CIS-TRANS ISOMERISM

INTRODUCTION

The best known form of *cis-trans* isomerism, that of ethylenic systems, was predicted by van't Hoff[3b] in 1875; the nomenclature (*cis* and *trans*) was introduced by Baeyer. In the classical period *cis-trans* isomerism was recognized in systems containing double bonds and also in certain alicyclic systems. The projected valency angles in the classical models of the two stereoisomeric 1,2-dimethyl-ethylenes and the two cyclohexane-1,4-dicarboxylic acids, for example, are 0 degrees for the *cis* and 180 degrees for the *trans* isomer's substituents. This characterization is equivalent to the classical definition which describes the isomers by reference to a plane in the molecule. Thus, 1,2-dimethylethylene has the two methyl groups of the *cis* isomer on the same side of a plane perpendicular to the plane of the molecule and cutting the ethylenic carbon atoms. The increased knowledge of molecular geometry has given the terms *cis* and *trans* a more general meaning and the terms are applied to any two non-optical stereoisomers in which two substituents have different projected valency angles. The projected valency angles can have any values θ_1 and θ_2 with $\theta_1 < \theta_2$; the *cis* isomer has the angle θ_1. A *cis* isomer having a projected valency angle of substituents of θ_1 has an intramolecular plane having the substituents on the same side; substituents in the *trans* isomer are on opposite sides of this plane. The determination of projected angles or the detection of a plane as a means of defining isomers may be simplified frequently since the substituents in a *cis* form are closer together than in a *trans* form and the separation of groups is usually easy to detect by inspection of a perspective diagram. The generality and value of the nomenclature is shown in its application, in modified form, to optical isomers; thus the terms cisoid and transoid are used as a means of further defining the spatial relationships of substituents in these isomers. As pointed out earlier π, *os*, *oc* and *r* types of *cis-trans* isomers may be recognized and this classification provides a basis for discussion.

o

π-TYPE *CIS-TRANS* ISOMERISM

ALIPHATIC MOLECULES

The structural units $>C=C<$, $>C=N-$ and $-N=N-$ have their atoms linked by a σ- and a π-bond. This bonding, in the absence of other factors, creates a planar, rigid structure and a high barrier to rotation exists. In an appropriately substituted ethylene two different permutations of substituents about the double bond are possible and these two geometries are separated by a high

rotational barrier. These *cis* and *trans* geometries are symmetrical, both isomers have a plane of symmetry and, in addition, the *trans* form can have a centre of symmetry if the substituents are identical and if the π-unit is a $>C=C<$ or $-N=N-$ type. The molecules (I)–(XII) are some examples of π-*cis-trans* isomers. The isomer pair (IX)–(X) presents a nomenclature problem and reference substituents must be chosen in order to define isomers. Thus, (IX) could be called *cis* if Br and I atoms are being referred to whereas referring to the Cl and I substituents the isomer would be defined as *trans*; unfortunately no general nomenclature is in use to define isomers such as (IX) and (X). The nomenclature problem is occasionally obviated since some isomers have trivial names, thus (XI) is called tiglic acid and (XII) is called angelic acid.

Oximes such as (XIII) and (XIV) are examples of *cis-trans* isomerism about a $>C=N-$ unit. The *cis-trans* isomerism of oximes is a matter of classical interest (Hantzsch and Werner[181]) and there

XIII	**XIV**	**XV**

has been much debate about its validity. Thus, it has been considered that the two oximes which may be isolated are structural isomers and not stereoisomers, however, the classical description of the oximes as *cis-trans* forms appears to be correct (Jerslev[182]). A nomenclature problem arises with oximes; the isomer (XIII) is called *cis* (OH and H closer together) and (XIV) is the *trans* form. An alternative nomenclature, of historical origin, is also in general use, the *cis* isomer is called syn and the *trans* isomer is called anti. An oxime of the type (XV) requires reference to R_1 or R_2 before it can be defined unambiguously as *cis* or *trans*. A variety of other systems have been obtained in *cis* and *trans* forms; phenylhydrazones $RCH = N \cdot NHC_6H_5$, semicarbazones $RCH = N \cdot NH_2CONH_2$ and Schiff's bases $RCH = NR'$ are some examples.

Cis-trans isomerism about an $-N=N-$ unit is illustrated by the azobenzenes (Hartley[183]); Azoxy compounds, $R-N=N(O)R'$, are also capable of *cis-trans* isomerism. A variety of other compounds

XVIII	**XIX**	**XX**	**XXI**

contain the $-N=N-$ unit notably the diazosulphonates and the diazocyanides and are expected to show *cis-trans* isomerism, however, the matter has been the subject of some controversy. Hantzsch[184] regarded the two forms of diazosulphonates as *cis* (XVIII) and *trans* (XIX) isomers. The diazocyanides were considered similarly to be the *cis* (XX) and *trans* (XXI) forms by Hantzsch and

Schultze[185]. The diazosulphonates were regarded as structural isomers by Bamberger and they were written as (XXII) and (XXIII), they differ in the constitution of the —SO₃K group and not in the geometry of the double bond; the diazocyanides were regarded by Orton as the structural isomers (XXIV) and (XXV).

$$R-N\!\!=\!\!N-SO_3K \qquad\qquad R-N\!\!=\!\!N-OSO_2K$$

XXII   **XXIII**

$$R-N\!\!=\!\!N-CN \qquad\qquad R-N\!\!=\!\!N-NC$$

XXIV   **XXV**

The structural isomerism theory of Bamberger[186] and Orton[187] was supported by Hodgson and Marsden[188] but more recent evidence (Freeman and Lefevre[189]) strongly supports the *cis-trans* isomerism interpretation of Hantzsch.

Systems containing more than one π-unit may give more than two *cis-trans* isomers; systems of this type may be conjugated, non-conjugated or cumulative. Conjugated and non-conjugated systems give a number of isomers which depends on the number of double bonds present and on the nature of the substitution. Thus, the three muconic acids (XXVI)–(XXVIII) are known, they are

XXVI   **XXVII**   **XXVIII**

defined as *cis-trans*, *trans-trans* and *cis-cis* respectively. The deduction of the number of primary isomers for substances such as muconic acid, by permutation, presents the problem of the geometry of the parent hydrocarbon. The problem is solved by assuming a staggered or *trans* structure throughout the chain of the parent hydrocarbon. Secondary isomers are possible, as will be shown later, but in general those isomers having a *trans* chain are most stable and so properly represent the primary forms. The naming of isomers such as (XXVI)–(XXVIII) requires a conventional starting point. thus.

starting from the upper —COOH group of (XXVI) it is clearly in a *cis* relationship to the substituent —CH=CH—COOH. The second —COOH group is clearly *trans* to the other substituent, thus, (XXVI) is defined as a *cis-trans* isomer. When the terminal groups are different the permutation method shows the existence of four primary isomers (XXIX)–(XXXII), they are named

trans-trans, cis-cis, trans-cis and cis-trans respectively. The isomer (XXXIII) has three double bonds and is called *trans-cis-trans*, more complex systems may be named similarly. The number of primary isomers for a symmetrically substituted conjugated system of n double bonds is given by the formula $2^{n-1} + 2^{p-1}$ where $n = 2p$ when n is even and $n = 2p - 1$ when n is odd (Kuhn and Winterstein[190]). Non-conjugated polyethenoid systems also have more than two stereoisomers but cumulative systems with an odd number of double bonds have only two *cis-trans* forms, for example (XXXIV)–(XXXV). The relative stabilities of *cis-trans* isomers have been discussed previously, non-bonded interaction, the electrostatic and resonance factors are involved and the *trans* form is usually the more stable though it rarely differs greatly in stability from the *cis* form.

ALICYCLIC MOLECULES

Alicyclic molecules may have an intra or extracyclic double bond; the isomers (XXXVI)–(XXXIX) are examples of *cis-trans* isomers having an extracyclic double bond, the isomer problem is clearly analogous to that of an aliphatic system. The *cis-trans* isomerism of molecules having an intracyclic double bond is rather more complex.

XXXVI   XXXVII

XXXVIII   XXXIX

Thus, cyclopropene (XL) consists of an ethylenic unit having two of its valencies joined to a methylene group, the methylene group must be attached to two *cis* ethylenic valencies since an attachment to the *trans* valencies produces a system which is too strained to exist. Cyclobutene, cyclopentene, cyclohexene and cycloheptene similarly

XL   XLI   XLII   XLIII

exist in only *cis* forms. However, in the case of cyclo-octene the ring size permits the existence of both *cis* (XLI) and *trans* (XLII) forms although in the latter some strain exists and the double bond is twisted slightly. Higher unsaturated alicyclic molecules can exist more easily as *cis* and *trans* forms and examples have been known for some time. Thus, civetone (XLIII) can exist in strain free *cis* and *trans* forms; a strain free *trans* cycloalkene probably occurs first with cyclononene.

AROMATIC MOLECULES

Aromatic molecules as such do not give π-isomers but aromatic groups may act as substituents in π-units, for example (III) and (IV). The conjugation of an aryl group with a π-unit reduces the height of the *cis-trans* overlap barrier.

σs-TYPE *CIS-TRANS* ISOMERISM

As mentioned earlier the existence of a steric barrier to rotation about a single bond can create *cis-trans* isomers.

ALIPHATIC MOLECULES

Aliphatic molecules, such as ethylene dibromide, which can be described as showing *cis-trans* isomerism have been discussed previously.

ALICYCLIC MOLECULES

The chair and boat forms of cyclohexane are *trans* and *cis* isomers respectively; the molecules (XLIV) and (XLV) are also σs-isomers.

XLIV   XLV

AROMATIC MOLECULES

Aromatic units are excellent for the creation of σs-type *cis* and *trans* isomers; appropriately substituted terphenyls, such as (XLVI) are examples (Stanley and Adams[191]). The parent hydrocarbon of (XLVI) can have the three geometries shown by the projections (XLVII)–(XLIX). These geometries arise because of steric inter-action of the type found in diphenyls; the geometries (XLVII) and (XLVIII), which are enantiomers correspond to a rotation of the end rings in opposite directions, the form (XLIX) is symmetrical having both a centre and a plane of symmetry. The permutation of the substituents on the parent molecules gives the different projections (L)–(LV). The pairs of enantiomers (L)–(LII) and (LI)–(LIII) are interconvertible by the rotation of the end rings through resonance barriers, similarly (LIV) may be converted to (L) or (LII) and (LV) may be converted to (LI) or (LIII) by a rotation through a resonance barrier. Thus, only two primary isomers exist and these equilibrate as the mixtures (L) ⇌ (LII) ⇌ (LIV) and (LI) ⇌ (LIII) ⇌ (LIV); these equilibrium mixtures consist of a pair of enantiomers and a symmetrical form and so are optically inactive. An examination of (L) and (LI) or (LII) and (LIII) or (LIV) and (LV) shows that in (L) and (LII) the X groups are

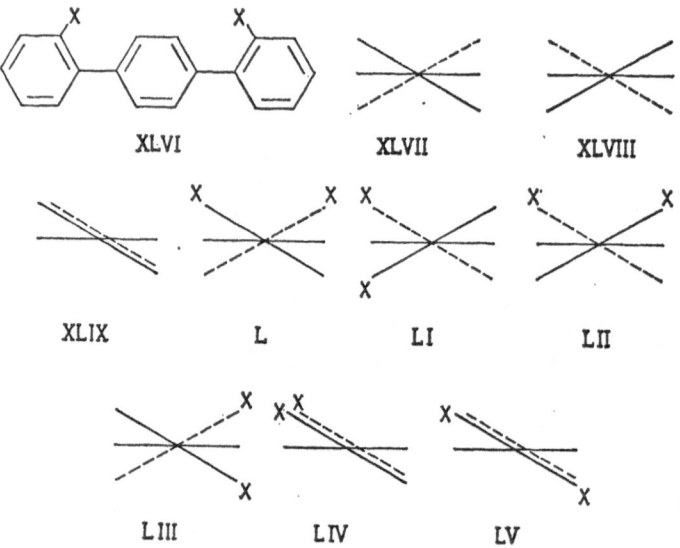

distant while in the equilibrated form (LIV) the substituents are close, however reference to the plane in which the central ring lies shows that the substituents in (L), (LII) and (LIV) lie on the same side of the plane and so this equilibrium mixture of isomers can be called the *cis* form. The mixture of (LI), (LIII) and (LV) makes up the *trans* form. It will be clear that the stability of *cis-cis-trans* isomers depends on the height of the steric barrier involved. In order to create primary *cis* and *trans* terphenyls it is actually necessary to substitute the central ring as would be expected from the discussion of the diphenyls. The substances (LVI) and (LVII) are terphenyls which have been obtained in primary *cis* and *trans* forms.

A variety of substituted mesitylenes of type (LVIII), R = $N(CH_3) \cdot SO_2 \cdot CH_3$ for example, has been obtained in *cis-trans* forms (Adams and Blomstrom[192]). The non-bonded interaction between the R groups and the adjacent nuclear methyl groups results in a steric barrier to rotation about the ring —R bond and a non-planar system results. The parent geometries of (LVIII) can be drawn as (LIX)–(LXI) and permutation shows a primary *cis* form derived from (LIX) and a pair of enantiomeric transoid forms derived from (LX) and (LXI).

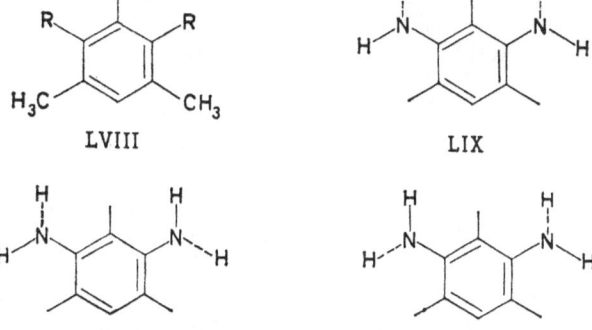

LVI·

LVII

LVIII

LIX

LX

LXI

σc-TYPE *CIS-TRANS* ISOMERISM

ALIPHATIC MOLECULES

Examples such as *cis* and *trans*-but-1,3-dienes have been mentioned previously. The existence of *σc*-isomers depends on the fact that a classical single bond in a conjugated system can acquire some double bond character. The central single bond in the muconic acids has some double bond character and a primary isomer such as (LXII) equilibrates with the secondary form (LXIII), thus conjugated systems show both π- and σc-isomerism.

LXII   **LXIII**

ALICYCLIC MOLECULES

The lactones and lactams are examples of alicyclic molecules showing σc-isomerism. The resonance barrier in these molecules arises from the conjugation of a lone pair on either oxygen or nitrogen with the carbonyl group and the CO—O or CO—NH bond acquires some double bond character. The lower lactones and lactoms, as expected, are found only in *cis* form but with higher members the *trans* form is known also; thus, the lactones (LXIV)

LXIV   **LXV**

exist as *cis* forms for $n = 4$–7 and as *trans* forms for $n = 10$–16, when $n = 8$ or 9 substantial amounts of both *cis* and *trans* isomers exist in equilibrium (Huisgen and Ott[193]). Lactams (LXV) exist as *cis* isomers for rings having up to nine members, for larger rings the *trans* form can exist and when $n = 7$ the *cis* form is most stable in solution but the crystal is made up of the *trans* isomer (Huisgen[194]).

AROMATIC MOLECULES

Some examples of this type have been mentioned previously; the pairs (LXVI)–(LXVII) and (LXVIII)–(LXIX) are aromatic *oc-cis-trans* isomers. The *cis* isomer (LXVIII) is stabilized by hydrogen bonding between the hydroxylic hydrogen and the chlorine atom. The secondary forms of some diphenyls and terphenyls are *oc*-isomers; *cis-trans* isomerism arising from a resonance barrier is always secondary in type.

| LXVI | LXVII | LXVIII | LXIX |

r-TYPE *CIS-TRANS* ISOMERISM

This type of *cis-trans* isomerism is restricted, of course, to alicyclic systems. The compound 1,3-dimethylcyclobutane has the two different molecular geometries (LXX) and (LXXI); these stereoisomers have no asymmetric centres, they are symmetrical and clearly

| LXX | LXXI | LXXII |

| LXXIII | LXXIV |

represent *cis* and *trans* isomers respectively. In a similar way all appropriately substituted even membered alicyclic rings are capable of *cis-trans* isomerism. Thus, the 1,4-dimethycyclohexanes and the cyclo-octane-1,5-dicarboxylic acids exist as *cis* and *trans* forms. The presence of heteroatoms may modify the conclusions applicable to carbocyclic rings, thus (LXXII) is not a *cis* isomer but (LXXIII) and (LXXIV) are *cis* and *trans* forms respectively. Conformational analysis does not affect the number of primary *cis-trans* isomers predicted by classical theory; the *cis* and *trans* forms of 1,4-dimethyl-cyclohexane are represented by the conformations (LXXV) and (LXXVI), these forms equilibrate with their inverse puckered

LXXV   LXXVI

conformations. *Cis-trans* isomers of the *r*-type are primary and can be interconverted only by a vibrational process.

Polycyclic species may exist as pure *cis* or as *cis* and *trans* forms depending on the sizes of the rings involved and on their mode of fusion. The compound cyclohexene oxide (LXXVII) is expected to

LXXVII   LXXVIII   LXXIX

$$CH_2 \text{---} CH \text{---} CH_2$$
$$CH_2$$
$$CH_2 \text{---} CH \text{---} CH_2$$

LXXX   LXXXI

have the three membered oxide ring *cis* fused, however, the cyclo-hexane ring has the chair form and this is considerably distorted when the oxide ring is *cis* fused (Ottar[195]). The *cis-trans* isomeric decalins (LXXVIII) and (LXXIX) have been discussed earlier; the rings are strainless. The molecule (LXXX) has the 'boat' structure (LXXXI) and is a *cis* form; a *trans* isomer would be too highly strained to exist.

10

THE CLASSIFICATION OF STEREOISOMERS RELATIVE AND ABSOLUTE STEREOCHEMISTRY

THE CLASSIFICATION OF STEREOISOMERS

Previous discussions have classified stereoisomers as optical or *cis-trans* and the designations given earlier to particular isomers are those most generally used. The terms optical and *cis-trans* are the best means of description but their use is by no means clear cut. Thus, at various points of earlier discussion ethylene dibromide has been described as showing *cis-trans* and optical isomerism and as having or not having asymmetric centres. The isomers (I)–(IV) also illustrate the nomenclature problem, thus

these isomers may be described as pairs of *cis-trans* isomers or, alternatively, as pairs of optical isomers. It is of obvious scientific value to have as unambiguous a means as possible for the designation of isomers and a method based on the general analysis for stereoisomer number provides this.

The determination of isomer number starts from the Kekulé-Couper structure which involves no geometrical features. An inspection of the Kekulé-Couper structure detects the parent hydrocarbon and this is then represented in hypothetical or actual geometrical form or forms. The permutation of substituents on the parent geometries then gives, as different permutations, the total number of stereoisomers for the substance. A consideration of possible internal rotations and vibrations then enables the different permutations to be grouped together as secondary forms. This procedure for deducing isomer number avoids the confusing issue of the asymmetric centre and simplifies classification.

The individual isomers are examined now for symmetry; isomers which are symmetrical can be designated as *cis* and *trans* by reference to some arbitrary molecular plane or by reference to the separation of two substituents. Isomers which are dissymmetric always occur as enantiomeric pairs and are called optical isomers.

V   VI

VII   VIII   IX

Thus, lactic acid derives from methane, as the parent hydrocarbon, and permutation of the substituents gives the two geometries (V) and (VI). These geometries are dissymmetric, they have a net interaction with polarized light, and so are described best as optical isomers. The tartaric acids derive from ethane and (VII)–(IX) represent their primary forms, the isomer (VII) is symmetrical and so is a *cis* or a *trans* form; reference to the plane cutting the ethane carbons and the carboxyl groups shows that the hydroxyl groups are on opposite sides hence the isomer is a *trans* form. Alternatively the hydroxyl (or other groups) are clearly as distant as is possible and the isomer is the *trans* form; the projected valency angle of the groups is 180 degrees. The isomers (VIII) and (IX) are dissymmetric and are classified as optical forms; these isomers may be defined further as cisoid since the hydroxyl groups are as close as possible and have a projected angle of 60 degrees. The rotational forms of the molecules (VII)–(IX) are all dissymmetric and so are optical isomers. In previous discussion the primary tartaric acid (VII) was considered to be an optical form and was referred to as the meso form; the present classification defines (VII) as a *trans* isomer and the term meso, which relates to symmetry properties, becomes unnecessary. However, the term meso is firmly established in the literature and perhaps (VII) is best described as meso-*trans* provided it is realized that the term meso is not connected, on the present classification, with optical isomerism. The cyclopentane

(X) is a *cis* isomer and the cyclohexane (XI) is an optical isomer. The conformational forms of 1,2-dimethylcyclohexane are all optical isomers whereas the two permutations for the cyclohexane-1,4-dicarboxylic acids, in Baeyer form, are *cis-trans* isomers. The classical forms of the inositols (XII)–(XX) may be described by considering the pairs of hydroxyl groups. Thus, considering (XII) and starting at C_1 the hydroxyl group pairs are described as *cis-cis-cis-cis-cis-cis*, the isomer (XIII) is *trans-cis-cis-cis-cis-trans* and (XII)–(XIII) may be regarded as a *cis-trans* pair since they differ only in the configuration of one hydroxyl group. The isomers (XIX) and (XX) are dissymmetric and constitute a pair of enantiomers.

Ethylenic substances such as the two 1,2-dimethylethylenes are clearly *cis* and *trans* forms; a pair of primary diphenyls are optical isomers as are allenes and spirans. Compounds such as ethylene dibromide are made up of a *trans* and two optical isomers; intramolecularly overcrowded species are optical isomers.

RELATIVE AND ABSOLUTE STEREOCHEMISTRY OF MOLECULES

A complete specification of the stereoisomerism of a substance requires a statement of the number and type of the primary and secondary forms together with the actual geometry of these forms. The actual or absolute stereochemistries of isomers have been determined by theoretical, physical or chemical methods or by a combination of these methods.

The isomers called maleic and fumaric acids have melting points of 130° and 286° respectively. Classical theory was able

XXI   XXII   XXIII

to say that these isomers are represented by the general geometries (XXI) and (XXII) but the theory alone was unable to decide whether (XXI) or (XXII) represented the actual geometry of the particular isomer maleic acid. Chemical examination shows that maleic acid, m.p. 130° forms, on melting, an anhydride which must have the structure (XXIII); the isomer m.p. of 286° does not form an anhydride under these conditions. The structure of the anhydride clearly requires the COOH groups to be close together and so maleic acid must have the geometry (XXI) and fumaric acid is (XXII). Results of the above type were obtained in the classical period and may be regarded as determinations of absolute stereochemistry since it was possible to determine the dispositions of the COOH groups and this is the salient stereochemical feature of the molecules. However, a rigorous determination of geometry for cases such as (XXI) and (XXII) had to await the development of theoretical and physical methods; thus, classical theory and experiment could show that the COOH groups of maleic acid are close together but only by modern physical methods has it been possible to determine how close together the groups are.

The assignm ent of absolute geometry to *cis-trans* isomers is rarely as simple as for the maleic and fumaric acids but nevertheless by appropriate physical and chemical methods it is usually possible to determine the absolute configuration of *cis* and *trans* forms. These methods depend on the physical and chemical consequences of molecular geometry and are beyond the scope of the present discussion. However, it is worth remembering that, in general, *trans* isomers are more stable, have higher melting points and are less soluble than *cis* forms, and this provides a simple though useful geometrical diagnosis. The absolute geometries of some common *cis-trans* isomers are shown as (XXIV)–(XXXVII). The absolute stereochemistries of many isomers of the above type were determined in the classical period; the formulation of strainless cyclohexane was introduced by Mohr but the confirmation of this detailed molecular geometry had to await the development of physical methods. Classical chemical, theoretical and physical methods were successful in determining the absolute geometry of many π- and r-*cis-trans* isomers but it proved impossible to determine the actual geometry of isomers such as meso-*trans*-tartaric acid and σs- and σc-isomers had not even been recognized. The isomer *trans*-tartaric acid was known to be symmetrical but it was not possible to decide whether it was a *cis* or a *trans* form and the principle of free rotation confused the issue further. Fischer arbitrarily adopted the *cis* geometry but chemical evidence was provided as early as 1931 (Amadori) to support the *trans* geometry. The problem has been resolved only recently and a variety of evidence, including that of conformational analysis, shows that meso-tartaric acid is a *trans* form. Conformational analysis is a valuable, though not unambiguous method of deciding the absolute geometry of meso forms.

The absolute stereochemistry of enantiomers presents a rather more difficult problem than that of *cis-trans* isomers and has been solved only recently. The problem of enantiomers is due to their close similarity and special physical methods had to be developed. The absolute geometry of the enantiomeric tartaric acids was determined by Bijvoet, Peerdeman and van Bommel[196] by a special x-ray method.

The dextro-rotatory enantiomer of tartaric acid is found to have the absolute geometry (XXXVIII), which happens to be the same

215

m.p. 15·5°

XXIV

m.p. 72°

XXV

m.p. 68°

XXVI

m.p. 133°

XXVII

m.p. 64°

XXVIII

m.p. 45°

XXIX

m.p. 155°

XXX

m.p. 170°

XXXI

m.p. 167°

XXXII

m.p. 309°

XXXIII

m.p. 125°

XXXIV

m.p. 6°

XXXV

m.p. 68°

XXXVI

m.p. 71°

XXXVII

as that attributed to it (in eclipsed form) by Fischer. More recently the absolute geometry of some enantiomeric diphenyls, which do not contain asymmetric centres, has been determined.

XXXVIII

XXXIX

XL

XLI

Prior to the modern determination of the absolute geometry of enantiomers the classical theory and experiment had been able to determine the relative configuration of quite complex enantiomers. Thus, Fischer was able to show that if the configuration of C_5 in (+)-glucose is assumed to be as shown in the projection (XXXIX) then the geometry of the other centres must also be as shown in (XXXIX). The actual geometry of C_5 was, of course unknown at the time and the hydroxyl group may well have been on the left in a projection representing the actual geometry. It follows that if the relative configuration of a molecule is known then the determination of the actual geometry of one asymmetric centre automatically determines the absolute geometry of all the other centres. The geometrical relationships between asymmetric centres in different molecules has also been extensively studied. Thus, (+)-glyceraldehyde, which was arbitrarily assigned the configuration (XL), has been converted into (−)-lactic acid which therefore must be represented by (XLI). Actually it is known now that (XL) represents the actual geometry of dextro-rotatory glyceraldehyde and so (XLI) represents the absolute geometry of (−)-lactic acid.

Glyceraldehyde is the simplest sugar molecule and may be related to more complex sugars, clearly the determination of the absolute geometry of the enantiomeric glyceraldehydes provides the absolute stereochemistry of the sugars. A large number of correlations of the above type have now been achieved and knowing the absolute geometry of an enantiomer provides a determination of the absolute geometry of all enantiomers which have been related to it.

BIBLIOGRAPHY

1 Kekule, *Liebigs Ann.* 106 (1858) 129
2 Couper, *Phil. Mag.* 16 (1858) 104
3 Van't Hoff, (a) *Voorstel tot Uitbreiding der Structuurformules in der Ruimte.* 1874 Utrecht; J. Greisen (b) *Bull. Soc. chim. Fr.* [2], 23 (1875) 295; (c) *The Arrangement of Atoms in Space.* 1898. London; Longmans
4 Le Bel, *Bull. Soc. chim. Fr.* [2], 22 (1874) 237.
5 Werner, (a) *Beiträge zur Theorie der Affinität und Valenz.* 1891; (b) Z. *anorg. Chem.* 3 (1893) 267, 294, 310; (c) *New Ideas on Inorganic Chemistry.* 1911. London; Longmans
6 Sidgwick and Powell, *Proc. Roy. Soc.* A 176 (1940) 153
7 Bohr, *The Theory of Spectra and Atomic Constitution.* 1922. Cambridge; Cambridge University Press
8 Stoner, *Phil. Mag.* 48 (1924) 719
9 Hund, *Linienspektren und periodisches System der Elemente.* 1927. Berlin; Julius Springer
10 Lewis, *J. Amer. Chem. Soc.* 38 (1916) 762
11 Pauli, Z. *Phys.* 31 (1925) 765
12 Schrödinger, *Ann. Phys. Lpz.* 79 (1926) 361
13 Pauling, (a) *Proc. nat. Acad. Sci., Wash.* 14 (1928) 359; (b) *J. Amer. chem. Soc.* 53 (1931) 1367; (c) *Nature of the Chemical Bond.* 1940. New York; Cornell University Press
14 Henry, *C. R.* 104 (1887) 1106
15 Erlenmeyer, Z. *Chem.* 7 (1862) 1
16 Baeyer, *Ber. Dt. chem. Ges.* 18 (1885) 2277
17 Ingold and Thorpe, *J. chem. Soc.* 115 (1919) 320; 117 (1920) 591
18 Penney, *Proc. Roy. Soc.* A 144 (1934) 166; A 146 223
19 Hückel, Z. *Phys.* 70 (1931) 204
20 Pauling and Sherman, *J. Amer. chem. Soc.* 59 (1937) 1450
21 Maccoll, *Trans. Faraday Soc.* 46 (1950) 369
22 Koch and Hammond, *J. Amer. chem. Soc.* 75 (1953) 3445
23 Zimmermann and Van Rysselberghe, *J. chem. Phys.* 17 (1949) 598
24 Bastiansen and Hassel, *Tidsskr. Kemi Bergv.* 6 (1946) 71
25 Coulson and Moffitt, *Phil. Mag.* 40 (1949) 1
26 Gomberg, *J. Amer. chem. Soc.* 22 (1900) 757
27 Bartlett and Knox, *J. Amer. chem. Soc.* 61 (1939) 3184
28 Franklin and Field, *J. chem. Phys.* 21 (1953) 550
29 Cram, Allinger and Langemann, *Chem. & Ind.* (1955) 919
30 Skell, Woodworth and McNamara, *J. Amer. chem. Soc.* 79 (1957) 1253
31 Mulliken, Reike and Brown, *J. Amer. chem. Soc.* 63 (1941) 41
32 Badger and Bauer, *J. chem. Phys.* 5 (1937) 599
33 Muller and Mulliken, *J. Amer. chem. Soc.* 80 (1958) 3489
34 Baker and Nathan, *J. chem. Soc.* (1936) 236
35 Thiele, *Liebigs Ann.* 306 (1899) 92
36 Walsh, *Proc. Roy. Soc.* A 174 (1940) 220; *Nature, Lond.* 157 (1940) 768

[37] Kekule, *Liebigs Ann.* 137 (1865) 158
[38] Lonsdale, *Proc. Roy. Soc.* A 123 (1929) 494
[39] Karle, *J. chem. Phys.* 20 (1952) 65
[40] Pauling and Sherman, *J. chem. Phys.* 1 (1933) 606
[41] Dewar, *Chem. Soc. Special Publ.* No. 12 (1958) 343
[42] Lippincott and Lord, *J. Amer. chem. Soc.* 68 (1946) 1868
[43] Hedberg quoted by Livingston, *Annu. Rev. phys. Chem.* 5 (1954) 395
[44] Longuet-Higgins and Salem, *Proc. Roy. Soc.* A 251 (1959) 172
[45] Coulson and Golebiewski, *Tetrahedron*, 11 (1960) 125
[46] Doering and Knox, *J. Amer. chem. Soc.* 76 (1954) 3203
[47] Schlenk and Holtz, *Ber. Dt. chem. Ges.* 49 (1916) 603
[48] Wepster, *Rec. Trav. chim. Pays-Bas* 72, (1953), 661
[49] Robinson, *J. chem. Soc.* (1916) 1038; 4th *Solvay Rep.* 1931
[50] Mills and Bain, *J. chem. Soc.* (1910) 1866
[51] Pope and Peachey, *J. chem. Soc.* (1899) 1127
[52] Mills and Warren, *J. chem. Soc.* (1925) 2507
[53] Ferriso and Hornig, *J. chem. Phys.* 23 (1955) 164
[54] Burrus and Gordy, *Phys. Rev.* 91 (1953) 313
[55] Harrison, Kenyon and Phillips, *J. chem. Soc.* (1925) 2552
[56] Phillips, Hunter and Sutton, *J. chem. Soc.* (1945) 146
[56a] Holliman and Mann, *Nature, Lond.* 159 (1947) 438
[57] Meisenheimer and Lichtenstadt, *Ber. Dt. chem. Ges.* 44 (1911) 365
[58] Hund, *Z. Phys.* 51 (1928) 759
[59] Lennard-Jones, *Trans. Faraday Soc.* 25 (1929) 668
[60] Mulliken, *Phys. Rev.* 32, 186 (1928) 761
[61] Fajans, *Z. phys. chem.* 130 (1927) 729
[62] Sidgwick, *The Electronic Theory of Valency.* 1927. Oxford; Oxford University Press
[63] Arndt, Scholz and Nachtwey, *Ber. Dt. chem, Ges.* 57 (1924) 1903
[64] Robinson, *An Outline on an Electrochemical Theory of the Course of Organic Reactions.* 1932. London; Institute of Chemistry
[65] Ingold, *Chem. Rev.* 15 (1934) 225
[66] Ingold, *J. chem. Soc.* (1933) 1120
[67] Schomaker and Stevenson, *J. Amer. chem. Soc.* 63 (1941) 37
[68] Coulson, *Proc. Roy. Soc.* A 169 (1937) 413
[69] London, *Trans. Faraday Soc.* 8 (1937) 33
[70] Born and Mayer, *Z. Phys.* 75 (1932) 1
[71] van der Waals, *Kontinuität des gas förmigen und flüssigen Zustandes.* 1873. Leiden; A. W. Sijthoff
[72] Stuart, *Z. phys. Chem.* B 27 (1927) 350
[73] Brown, *Science,* 103 (1946) 385
[74] Kistiakowski, Lacher and Stitt, *J. chem. Phys.* 7 (1939) 289
[75] Pitzer, *Disc. Faraday Soc.* 10 (1951) 66
[76] Eyring, *J. Amer. chem. Soc.* 54 (1932) 3191
[77] Pauling, *Proc. nat. Acad. Sci., Wash.* 44 (1958) 211
[78] Eyring, Stewart and Smith, *Proc. nat. Acad. Sci., Wash.* 44 (1958) 259
[79] Haworth, *The Constitution of Sugars.* 1929. London; Arnold
[80] Pitzer, *Chem. Rev.* 27 (1940) 39
[81] Aston, Schumann, Fink and Doty, *J. Amer. chem. Soc.* 63 (1941) 2029
[82] Johnson, Margrave, Bauer, Frisch, Dreger and Hubbard, *J Amer. chem. Soc.* 80 (1960) 1255

[83] Hassel, *Tidsskr. Kemi Bergv.* 3 (1943) 32
[84] Sachse, *Ber. Dt. chem. Ges.* 23 (1890) 1363
[85] Mohr, *J. prakt. Chem.* 2, 98 (1918) 815
[86] Prelog, *J. chem. Soc.* (1950) 420
[87] Kohlrausch, Reitz and Stockmair, *Z. phys. Chem.* B 32 (1936) 229
[88] Beckett, Freeman and Pitzer, *J. Amer. chem. Soc.* 70 (1948) 4227
[89] Wightman, *J. chem. Soc.* (1925) 1421
[90] Turner, *J. Amer. chem. Soc.* 74 (1952) 2118
[91] Bastiansen and Hassel, *Nature, Lond.* 157 (1946) 765
[92] Blomquist, Liu and Bohrer, *J. Amer. chem. Soc.* 74 (1952) 3643
[93] Ruzicka, Hurbin and Boekenoogen, *Helv. chem. Acta.* 16 (1933) 498
[94] Mulliken and Roothan, *Chem. Rev.* 41 (1947) 203
[95] Dewar, *J. Amer. chem. Soc.* 74 (1952) 3341
[96] Edwards and Meacock, *J. chem. Soc.* (1957) 2000, 2007, 2009
[97] Arnett and Wu, *Chem. & Ind.* (1959) 1488
[98] Donaldson and Robertson, *J. chem. Soc.* (1953) 17
[99] Bell and Waring, *J. chem. Soc.* (1949) 2689
[100] Herbstein and Schmidt, *J. chem. Soc.* (1954) 3302
[101] McIntosh, Robertson and Vand, *J. chem. Soc.* (1954) 1661
[102] Newman, Lutz and Lednicer, *J. Amer. chem. Soc.* 77 (1955) 3420
[103] Flurscheim, *J. prakt. Chem.* 66 (1902) 321
[104] Lewis, *Valency and the Structure of Atoms and Molecules.* 1923. New York; Chemical Catalog Coy. Inc.
[105] Stieglitz, *Proc. nat. Acad. Sci., Wash.* 1 (1915) 196
[106] Mulliken, *J. chem. Phys.* 2 (1934) 782; 3 (1935) 573
[107] Gordy, *J. chem. Phys.* 14 (1946) 314
[108] Malone, *J. chem. Phys.* 1 (1933) 197
[109] Keesom, *Z. Phys.* 22 (1921) 129, 643
[110] Lassettre and Dean, *J. chem. Phys.* 16 (1948) 157; 17 (1949) 317
[111] Gwinn and Pitzer, *J. chem. Phys.* 16 (1948) 303
[112] Von Schleyer, Trifan and Bacskai, *J. Amer. chem. Soc.* 80 (1956) 6691
[113] Meisenheimer, Angermann, Finn and Vieweg, *Ber. Dt. chem. Ges.* 57 (1924) 1747
[114] Barker, *Phys. Rev.* 33 (1929) 648
[115] Wall and Glockler, *J. chem. Phys.* 5 (1937) 314
[116] Manning, *J. chem. Phys.* 3 (1935) 136
[117] Cleeton and Williams, *Phys. Rev.* 45 (1934) 234
[118] Werner, *Lehrbuch der Stereochemie.* 1904. Jena; G. Fischer
[119] Hund, *Z. Phys.* 43 (1927) 805
[120] Debye, *Z. Phys.* 21 (1920) 178
[121] Newman, *J. Amer. chem. Soc.* 72 (1950) 4783
[122] Palin and Powell, *J. chem. Soc.* (1947) 208
[123] Schlenk, *Liebigs Ann.* 204 (1949) 565
[124] Smith, *Acta cryst.* 5 (1952) 224
[125] Meyer, *Ber. Dt. chem. Ges.* 23 (1890) 568
[126] Bell, Jones and Whiting, *J. chem. Soc.* (1957) 2597
[127] Lanpher, *J. Amer. chem. Soc.* 79 (1957) 5578
[128] Mulliken, *Rev. Mod. Phys.* 14 (1942) 265
[129] Wislicenus, *Liebigs Ann.* 246, 53
[130] Kincaid and Henriques, *J. Amer. chem. Soc.* 62 (1940) 1474
[131] Pasteur, *Ann. Chim.* (*Phys.*) 3, 24 (1848) 442

[132] Wislicenus, *Liebigs Ann.* (1873) 167 345
[133] Mohr, *J. prakt. Chem.* 2, 68 (1903) 369
[134] Aschan, *Ber. Dt. chem. Ges.* 35 (1902) 3389
[135] McCasland, Horvath and Roth, *J. Amer. chem. Soc.* 81 (1959) 2399
[136] Eliel, *J. Amer. chem. Soc.* 71 (1949) 3970
[137] Alexander and Pinkus, *J. Amer. chem. Soc.* 71 (1949) 1786
[138] Alexander, *J. Amer. chem. Soc.* 72 (1950) 3796
[139] Stewart and Allen, *J. Amer. chem. Soc.* 54 (1932) 4027
[140] Meisenheimer, *Ber. Dt. chem. Ges.* 41 (1908) 3966
[141] Meisenheimer, *Liebigs Ann.* 385 (1911) 117; 449 (1926) 191
[142] Kumli, McEwen and Vanderwerf, *J. Amer. chem. Soc.* 81 (1959) 249
[143] Pope and Peachey, *J. chem. Soc.* 77 (1900) 1072
[144] Harrison, Kenyon and Phillips, *J. chem. Soc.* (1926) 2079
[145] See Hudson, *Advanc. Carbohyd. Chem.* 3 (1948) 1
[146] Newman, *J. chem. Educ.* 32 (1955) 344
[147] Cahn, Ingold and Prelog, *Experientia*, 12 (1956) 81
[148] Amadori, *Gazz. chim. ital.* 61 (1931) 215
[149] Drew and Haworth, *J. chem. Soc.* (1926) 2303
[150] Bottini and Roberts, *J. Amer. chem. Soc.* 80 (1958) 5203
[151] Haresnape, *Chem. & Ind.* (1953) 1091
[151a] Angyal and McHugh, *Chem. & Ind.* (1956) 1147
[152] Tanret, *Bull. Soc. chim. Paris*, 3, 13 (1895) 728
[153] Mark, *Chem. Rev.* 26 (1940) 169
[154] MacDonald and Beevers, *Acta Cryst.* 5 (1952) 645
[155] Linstead, *Chem. & Ind.* 56 (1937) 510
[156] Hückel, *Liebigs Ann.* 441 (1925) 1; 451 109
[157] Turner and Lefevre, *Chem. & Ind.* 831 (1926) 883
[158] Bell and Kenyon, *Chem. & Ind.* 864 (1926)
[159] Mills, *Chem. & Ind.* 884 (1926) 905
[160] Mills and Maitland, *J. chem. Soc.* (1936) 987
[161] Pope, Perkin and Wallach, *J. chem. Soc.* (1909) 1789
[162] Kuhn and Wallenfels, *Ber. Dt. chem. Ges.* 71 (1938) 783
[163] Jansen and Pope, *Chem. & Ind.* (1932) 316
[164] Christie and Kenner, *J. chem. Soc.* (1922) 614
[165] Newman and Hussey, *J. Amer. chem. Soc.* 69 (1947) 3023
[165a] Adams, see *Organic Chemistry*. Gilman (Ed.) 1938. New York; Wiley
[166] Bastiansen, *Acta Chem. scand.* 4 (1950) 926
[167] Lesslie and Turner, *J. chem. Soc.* (1932) 2394
[168] Shaw and Turner, *J. chem. Soc.* (1933) 135
[169] Lesslie and Turner, *J. chem. Soc.* (1933) 1588
[170] Stanley and Adams, *J. Amer. chem. Soc.* 52 (1930) 1200
[170a] Lesslie and Turner, *J. chem. Soc.* (1932) 2021
[171] Hill, *J. chem. Phys.* 14 (1946) 465; 16 (1948) 399
[172] Westheimer, and Mayer, *J. chem. Phys.* 14 (1946) 733
[173] Calvin, *J. org. Chem.* 4 (1939) 256
[174] Cagle and Eyring, *J. Amer. chem. Soc.* 73 (1951) 5628
[175] Ahmed and Hall, *J. chem. Soc.* (1959) 3383
[176] Mills and Elliott, *J. chem. Soc.* (1928) 1291
[177] Theilaker and Baxmann, *Liebigs Ann.* 581 (1953) 11
[178] Newman, *J. Amer. chem. Soc.* 62 (1940) 2295
[179] Wittig and Zimmermann, *Ber. Dt. chem. Ges.* 86 (1953) 629

BIBLIOGRAPHY

[180] Robinson, *Proc. Roy. Soc.* A 192 (1948) 14

[181] Hantzsch, and Werner, *Ber. Dt. chem. Ges.* 23 (1890) 11

[182] Jerslev, *Nature, Lond.* 280 (1957) 1410

[183] Hartley, *J. chem. Soc.* (1938) 633

[184] Hantzsch, *Ber. Dt. chem. Ges.* 27 (1894) 1715, 1726

[185] Hantzsch and Schultze, *Ber. Dt. chem. Ges.* 28 (1895) 2073

[186] Bamberger, *Ber. Dt. chem. Ges.* 27 (1894) 2586, 2930

[187] Orton, *J. chem. Soc.* (1903) 805

[188] Hodgson and Marsden, *J. chem. Soc.* (1943) 470

[189] Freeman and Lefevre, *J. chem. Soc.* (1951) 415

[190] Kuhn and Winterstein, *Helv. chim. Acta.* 11 (1928) 87

[191] Stanley and Adams, *Rec. Trav. chim. Pays-Bas* 48 (1929) 1035

[192] Adams and Blomstrom, *J. Amer. chem. Soc.* 75 (1953) 2375

[193] Huisgen and Ott, *Tetrahedron,* 6 (1959) 253

[194] Huisgen, Brade, Walz and Glogger, *Ber. Dt. chem. Ges.* 69 (1957) 341

[195] Ottar, *Acta Chem. scand.* 1 (1947) 283

[196] Bijvoet, Peerdeman, and van Bommel, *Nature, Lond.* 168 (1951) 271

INDEX

Acetylene,
  bond lengths in, 12
  bond strengths in, 12
Alicyclic molecules, electrostatic effects in, 81
Aliphatic ethers, geometry of, 34
Allene, geometry of, 30
Allyl carbonium ion, 29
Amido group, geometry of, 32
Amines, geometry of, 31
Ammonia, geometry of, 31
Aniline, geometry of, 31
Aromatic molecules, conformational theory of, 71
Asymmetric centre,
  concept of, 113
  detection of, 113
  group geometry effect, and, 116
  isotope effect, and, 116
Atom,
  carbon,
    sp^3 3, 6
    sp^2, 7
    sp, 11
    $sp^{4.12}$, 16
    tricovalent, 16
  nitrogen,
    p^3, 31
    sp, 32
    sp^2, 32
    sp^3, 31, 33
  oxygen,
    p^2, 34
    sp^2, 35
    sp^3, 34, 35
  phosphorus, geometry of, 38
  sulphur, geometry of, 35
Atomic states, combinations of, 18
Azomethine group, geometry of, 32

Benzene, geometry of, 25
Benzyl,
  carbonium ion, geometry of, 29
  radical, geometry of, 30
Benzyne, geometry of, 29
Bond,
  axial, 61
  bent, 8

Bond—*cont.*
  covalent, energy of, 5
  double, rigidity of, 11
  equatorial, 61
  hydrogen, 82
  order, 46
  π, 9
  σ, 9
Born forces, 47
n-Butane, conformational theory of, 52
But-1,3-diene, geometry of, 21, 69

Carbanion, geometry of, 17
Carbon radical, geometry of, 18
Carbonium ion, geometry of, 17
Carbonyl group, geometry of, 35
Chloroacetyl chloride, geometry of, 81
Cis-trans,
  isomerism, 99, 199
  isomers, classification of, 101
Clathrate compounds, 96
Compression energy, 49
Conjugated systems, 21
Covalent radii, 45, 46
Cyano group, geometry of, 33
Cyclobut-1,3-diene, geometry of, 27
Cycloheptatriene, geometry of, 27
Cyclohexane, conformational theory of, 58
Cyclohexanes, substituted, conformational theory of, 61
Cyclohexanone, conformational theory of, 65
Cyclohexene, conformational theory of, 65
Cyclo-octatetrene, geometry of, 27
Cyclopentane, conformational theory of, 57
Cyclopentadiene, geometry of, 27
Cyclopentadienyl anion, geometry of, 32
Cyclopropane, geometry of, 15

Debye forces, 92
Decalins, geometry of, 63
Delocalization energy, 25
Diacetyl, conformation of, 81

225